EVIDENCE-INFORMED APPROACHES FOR MANAGING DEMENTIA TRANSITIONS

EVIDENCE-INFORMED APPROACHES FOR MANAGING DEMENTIA TRANSITIONS

Riding the Waves

Edited by

LINDA GARCIA

LIFE Research Institute, University of Ottawa,
Ottawa, ON, Canada
&
Interdisciplinary School of Health Sciences, Faculty of Health Sciences, University of Ottawa,
Ottawa, ON, Canada

LYNN MCCLEARY

Department of Nursing, Brock University, St. Catharines, Ontario, Canada

NEIL DRUMMOND

Alberta Health Services Chair in Primary Care Research, Department of Family Medicine,
University of Alberta, Canada

Academic Press is an imprint of Elsevier
125 London Wall, London EC2Y 5AS, United Kingdom
525 B Street, Suite 1650, San Diego, CA 92101, United States
50 Hampshire Street, 5th Floor, Cambridge, MA 02139, United States
The Boulevard, Langford Lane, Kidlington, Oxford OX5 1GB, United Kingdom

Library of Congress Cataloging-in-Publication Data
A catalog record for this book is available from the Library of Congress

British Library Cataloguing-in-Publication Data
A catalogue record for this book is available from the British Library

ISBN: 978-0-12-817566-8

For information on all Academic Press publications visit our website at https://www.elsevier.com/books-and-journals

Publisher: Andre Gerhard Wolff
Acquisitions Editor: Mary Preap
Editorial Project Manager: Sara Pianavilla
Production Project Manager: Sreejith Viswanathan
Cover Designer: Mark Rogers

Typeset by TNQ Technologies

Contents

Contributors

Melanie Deist
Department of Psychology, Stellenbosch University, Stellenbosch, South Africa

Bonnie M. Dobbs
Department of Family Medicine, University of Alberta, Edmonton, AB, Canada

Neil Drummond
Alberta Health Services Chair in Primary Care Research, Department of Family Medicine, University of Alberta, Canada

Willian Dullius
Escola Estadual de Ensino Médio Cônego João Batista Sorg - Government of the Rio Grande do Sul, Carazinho, Rio Grande do Sul, Brazil

Linda Garcia
Life Research Institute, University of Ottawa, Ottawa, ON, Canada; Interdisciplinary School of Health Sciences, Faculty of Health Sciences, University of Ottawa, Ottawa, ON, Canada

Abraham P. Greeff
Department of Psychology, Stellenbosch University, Stellenbosch, South Africa

Sharon Koehn
Gerontology, Simon Fraser University, Vancouver, BC, Canada; Sharon Koehn Research Consulting, Vancouver, BC, Canada

Geneviève Lemay
Faculty of Medicine, Division of Geriatrics, University of Ottawa, The Ottawa Hospital, Ottawa, ON, Canada; Montfort Hospital/Institut du Savoir Montfort Ottawa, ON, Canada

Peter A. Lichtenberg
Institute of Gerontology and Merrill Palmer Skillman Institute, Wayne State University, Detroit, MI, United States

Lynn McCleary
Department of Nursing, Brock University, St. Catharines, Ontario, Canada

Katherine S. McGilton
Research, Toronto Rehabilitation Institute- UHN, Toronto, ON, Canada; Lawrence S. Bloomberg, Faculty of Nursing, University of Toronto, Toronto, ON, Canada

Mackenzie Powell
Quinte Health Care, Belleville, Ontario, Canada

Muncuran Purewal
Department of Family Medicine, Faculty of Medicine, University of Ottawa, Ottawa, Canada

Annie Robitaille
Interdisciplinary School of Health Sciences, University of Ottawa, Ottawa, Ontario, Canada

Genevieve Thompson
Associate Professor, Nursing, University of Manitoba, Winnipeg, MB, Canada

Abigail Wickson-Griffiths
Assistant Professor, Nursing, University of Regina, Regina, SK, Canada

Stacey Wood
Scripps College, Claremont, CA, United States

Foreword

The ground work for this book derived from a large program of epidemiological research to which many of the Canadian authors contributed. The purpose of the research was to understand the impact of transitions on the health status and care processes of people with dementia and their caregivers. It was a substantial and long-term study which incorporated longitudinal and cross-sectional data collection, including quantitative and qualitative methods. We identified a number of common and significant transitions experienced by people with dementia and their caregiving family and friends, and we maintained regular contact with those in our study over several years.

This book is intended to consider the practical real-life application of the evidence and knowledge gained during our study and from other research about transitions experienced by people with dementia. It is intended primarily for service providers and policymakers but is about those who live with dementia and how to help them live and thrive in the face of the significant life changes the dementia might have caused. It is based on North American and European experiences, but it acknowledges diversity within those contexts. For this reason, we think it has wider relevance. Essentially, we have tried to translate our research into useful guidance, advice, or support for people similar to our research participants to make the navigation of transitions and other aspects of the dementia trajectory easier and more successful for everyone. We think of it as an example of "science meeting lived experience," hopefully to the benefit of both. Certainly, the science benefitted.

During the very early planning for that study, when most of us did not know each other very well, sitting round a big open square table in the Banff Centre in Alberta, surrounded by the Canadian Rocky Mountains, one of us asked the group "How many of us are caring for someone close to us who has dementia?" There were about 20 people in the room, and 19 of us put up our hands. The one who did not said "Are now, or have done?" The response was "Either or both." The last hand went up. This has always struck each of us as an important feature of our group. We are academicians, researchers, or clinicians, and some of us are all of those, but each of us have a personal stake in what we do. That includes this book.

The title derives from discussion in relation to what we would hope for if we experience dementia again, ourselves or with a family member or friend. Our hopes centre around being able to continue doing things that are meaningful to us, maybe in a different way. We hope for continued pleasure in activities related to our hobbies and pastimes, and we hope that we can savour the moments as we engage with those with whom we are close. If we are lucky, the paddlers among us will be able to run our favorite rivers and each run will become, for us, a challenging and thrilling first descent, no matter how often we do it. This hope suggested the title itself. It derived from some reflections among fellow paddlers. We liken the typical dementia trajectory to a whitewater river. There are fast sections and slower ones, drops and pools, and sometimes long areas of still water where nothing much seems to be happening at all. Getting down it in a canoe requires effort, expertise, knowledge, luck, and usually a bit of help. But if you can ride the waves, you can get down safely, enjoy at least some of it, and be changed by it. Eventually you may get to the point where you need a guide. It can still be meaningful and fun. We hope this book helps prolong a sense of positive experience in people with dementia and in those trying to help them.

ND, LG, LM

Acknowledgments

Much of this book draws heavily on the work of the pan-Canadian DementiaNet research group and on two of its studies in particular. The Dementia Transitions Study was funded by the Canadian Institutes of Health Research through the Partnerships for Health Care Improvement program (file number 91286), and the Pathways to Dementia Diagnosis Study was funded by the Canadian Social Science and Humanities Research Council. Over the years, a large team of investigators, research staff, and graduate students contributed to the study.

Above all, we thank all the research participants who generously shared their experiences with the research teams whose work features in this book and without whose contribution none of it would have been written.

CHAPTER 1

Improving the lived experience of dementia transitions

Neil Drummond[1], Linda Garcia[2,3], Lynn McCleary[4]
[1]Alberta Health Services Chair in Primary Care Research, Department of Family Medicine, University of Alberta, Canada; [2]LIFE Research Institute, University of Ottawa, Ottawa, ON, Canada; [3]Interdisciplinary School of Health Sciences, Faculty of Health Sciences, University of Ottawa, Ottawa, ON, Canada; [4]Department of Nursing, Brock University, St. Catharines, Ontario, Canada

Dementia is a progressive condition characterized by loss of memory and other cognitive functions. But its most significant impact on those who live with the condition is how it progressively affects daily activities. In the later stages it may affect movement, speech, and language and other noncognitive systems. The causes of these are still unclear, but the associated changes in the brain are physiological and associated with cell damage.

"Dementia" as a term refers to a syndrome of disease with symptoms of memory, cognitive, behavioral, language, and locomotive decline and, eventually, death. Whether you are a person with dementia or are close to someone who has it, it is a human and humbling experience. Alzheimer Disease International estimates that throughout the world, a new case of dementia develops every 3 seconds, although not all will be diagnosed [1]. The incidence of dementia rises exponentially between 65 and 90 years of age. Among those older than 65 years, around 7%–8% have dementia [2–4]. Among those older than 75 years, about 30% have the condition, and among those older than 85 years, over two-thirds of them do. Around 14% of all deaths may be attributed to dementia. Five-year mortality among people with dementia is 70%, compared with 35% in those without dementia [5]. Prevalence estimates for the most common individual types of dementia as proportions of the total number of dementia cases include 60% for Alzheimer's disease, 20% for vascular dementia, 15% for dementia with Lewy bodies, and 5% for frontotemporal dementia (also known as Pick's disease) [6].

Studies into dementia risk factors have shown that people with high risk for cardiovascular disease are also at high risk of dementia, although the relationship is not well identified nor understood [7]. Modifiable cardiovascular risk factors (physical inactivity, smoking, midlife hypertension, midlife obesity, diabetes), together with depression and low educational

Evidence-informed Approaches for Managing Dementia Transitions
ISBN 978-0-12-817566-8
https://doi.org/10.1016/B978-0-12-817566-8.00001-2

attainment, are associated with approximately 33% of patients with Alzheimer's disease [8]. Hypertension, hyperlipidemia, and type II diabetes are associated with an increase in the incidence of diagnosed dementia [9,10]. It has been recommended that maintaining appropriate body weight, blood pressure, blood glucose, and cholesterol levels might be causally linked to decreasing, or delaying, the incidence of dementia [11–13]. A 10% reduction in hypertension, diabetes, smoking, and other risks might reduce the prevalence of dementia by 8.3% [14]. A 1-year delay in the mean age of dementia onset could lead to a 10% reduction in dementia prevalence by 2050 [15]. Recent research suggests that the incidence of dementia is declining, along with a declining incidence of stroke, which might further suggest an association between cardiovascular disease risk control and overall health status in the community [16,17].

The onset of dementia is insidious and its development is normally very slow. As discussed in Chapter 2, diagnosis may not occur for a substantial period, during which the individual who has the disease and their caregivers struggle to explain slowly worsening cognitive function until a point is reached at which help is sought. Even then, diagnosis may be delayed [2,18]. Average life expectancy from the point of diagnosis is estimated at 8–10 years but is subject to much variation associated with age at onset, preceding general health status, comorbidity, and other factors [19]. Most people with dementia die from other causes. Alzheimer disease was the sixth most common cause of death in Canada in 2017, at 18/100,000 people [20].

Treatment for dementia typically involves a gradual increase in support for those with the disease and their caregivers. These supports include clinical and social services, initially in the community, later including respite care, and, eventually, long-term residential care. While dementia treatment does not require acute care or hospitalization, people with dementia are often admitted to hospital for treatment of other medical conditions, as discussed in Chapter 5.

A focus in research is finding a cure, but none are yet in sight. Pharmaceutical treatment depends on the type of dementia. Treatment usually consists of cholinesterase inhibitor medication intended to arrest cognitive decline, if not to reverse it. This medication is recommended for treatment of Alzheimer disease, Alzheimer disease mixed with vascular disease, and for dementia associated with Parkinson disease [21]. Evidence for use with vascular dementia is limited. There are no approved medications for the treatment of Lewy bodies dementia or frontotemporal dementia. Data on effectiveness of cholinesterase inhibitors for Alzheimer disease [22] indicated numbers needed to treat of seven people to achieve arrested decline in 1 person, 12 for slight cognitive improvement, and 42 for major cognitive improvement. The number needed to harm (mainly through side effects) was also 7. Later trials have been more optimistic, indicating moderate improvements in cognition and activities of daily living lasting six to 12 months in mild-to-moderate cases of dementia who had been previously prescribed the medication and continued on it, compared with those whose medication was discontinued [23].

As with many chronic health conditions, those who interact with people with dementia, as well as those who care for them, are also impacted by its effects. Contributions and experiences of caregivers are crucial to good quality of life. There is evidence that caregivers may obtain emotional or psychological satisfaction from caregiving despite the significant health hazards and burdens involved [24]. In general, both caregivers and people with dementia prefer that the person lives at home for as long as possible [25]. Interest in maintaining an acceptable quality of life for as long as possible is a key feature for those living with the disease. The willingness among caregivers to undertake a major role in the provision of care is too often taken for granted by formal service providers and funders. Yet, without their contribution, the circumstances and condition of those with dementia would be very much worse. In a book intended to help people navigate their way through problems commonly encountered in relation to dementia, we are constantly aware that it is both the person with dementia and their caregiver who require that help, even if the primary audience for this book are professional providers.

Caring, caregiving, care providing, care partnering (The terminology varies across the globe and through time), is an enormous aspect of the lived experience of dementia, for both the person with the condition and those doing the caring. The term denotes the activity (and often emotional commitment) of individuals, usually spouses, adult children, other family members, friends, or neighbors who undertake work to ensure that the person with dementia is "looked after." Caregiving is usually informal in the sense of there being no contractual obligation. It is usually unpaid. It is often "live-in." It is a source of enormous stress, anxiety, and fatigue, but it is also associated with a sense of responsibility being fulfilled, of commitment to a longstanding relationship, a duty to a valued individual. It is something we do only partly because we have to. We also do it for love. We often do it for far longer than we probably should, and we stop doing it with a combined sense of grief as well as relief. For these reasons, this book will continually focus on the experiences of caregivers as much as on those of individuals with the disease because they are crucial to achieving better health outcomes, effective processes, and positive patient experiences. And they are often overlooked, discounted, and misunderstood. We shall argue that caregivers and caregiving should be recognized as being as important a component in the health system for people with dementia as the professional care providers (physicians, nurses, social workers, dieticians, pharmacists, etc.) who get paid to do it. The idea of "riding the waves" is not just about patient experience. It is something shared by everyone who has significant contact or an emotional stake in the well-being of those individuals with the condition. The relationship is so codependent that the very demarcation between giving and receiving care can be blurred, with caregivers stating that they sometimes feel they receive care from the care recipient.

Recent thinking has identified "resilience" as a key concept related to successful health experiences, even in the face of significant illness. Part of this is about adjusting to different circumstances and conditions, implying a certain inherent toughness of character or personality. Adjustment has been identified as an important contributor to self-perceived quality of life. Social capital may also be implicated in relation to resilience, in the sense of providing a resource to deploy when overcoming problems, whether it derives from one's reputation or from one's social network. It is probably also about optimism and being positive in adversity as a personality trait. Hence, resilience is a quality deriving from multiple aspects of the self, of the social and physical environment, and of the nature of the problem, all of which are in an almost constant state of evolution, transition, and variation. We shall develop the idea of resilience throughout the book, particularly in relation to important transition events.

The increase in the prevalence of dementia globally [26,27] and the nature of the condition itself [28] demand that dementia services must very often involve interdisciplinary and multiple-agency collaboration and coordination [29], with the involvement of families [30], community-based services, and care facilities [31]. However, it has proven challenging to achieve and maintain collaboration and coordinated services as the symptoms of dementia progress and the needs of the person with dementia and their families change. The complexity of the condition often leads to people encountering a bewildering array of organizational and functional boundaries while attempting to navigate the transitions between services.

Despite attempting to address needs, this service complexity often, in practice, causes gaps in the care continuum [32].

Transition is defined by Webster's New Collegiate Dictionary as a "passage from one state, stage, subject or place to another" [33]. This includes periods of change in life phase, situation, or status between two comparatively stable periods. Transitions are precipitated by an event or turning point requiring a response and conclude when stability and equilibrium are achieved [34]. A transition period may lead to disequilibrium and upheaval, necessitating the development of new skills, behaviors, relationships, and strategies. People in transition may be more vulnerable to increased risk of illness or maladaptive coping [35].

Health-related transitions may be broadly classified according to whether they relate to health or illness itself, such as changed ability to function associated with progress of dementia; to a situation, such as a person with dementia moving from their home to live with a family member; or to a stage of physical or personal development, such as retirement of a caregiver [36]. Individuals typically experience several types of transitions simultaneously. For example, as discussed in Chapter 5, a person may be diagnosed with dementia when admitted to hospital for treatment of another illness, which might precipitate changes in family caregiving responsibilities and stimulate a need to start receiving support services at home, four simultaneous transitions. For this reason it has been suggested that patterns of response to numerous transitions should be studied, rather than responses to individual ones [35]. Furthermore, since transitions are processes which occur over a period of time, research designs need to capture the evolution of the transition experience [37]. On the basis of our review of the literature, our experience as clinicians, and also as past and present caregivers for family members with dementia, we identified the following transitions as both common, often very significant for persons with dementia, caregivers, and formal care providers, and appropriate for detailed study:

1. initial problem identification and diagnosis;
2. requiring support from external agencies;
3. driving cessation;
4. changes in financial autonomy;
5. acute hospital admission;
6. changes in informal support;
7. moving to new community-based living accommodation;
8. moving to long-term care;
9. entering palliative or end-of-life care.

The research on the lived experiences associated with these transitions is variable. Some have been looked at extensively (e.g., driving cessation and transition to long-term care), while others have had little attention (e.g., change in informal support). All may affect the well-being of people with dementia as well as those who provide care for them. Initial problem identification and help-seeking, contact with health services, diagnostic processes, disclosure of a diagnosis, and postdiagnostic experiences represent a set of related aspects which cluster around the first experiential transition of the dementia trajectory [38—44]. Studies report with some consistency that help seeking after the recognition of a problem is delayed, with many people being ambivalent about seeking or accepting assistance. There is evidence to suggest that perceived access may influence the decision to seek, or not to seek, help. Yet contemporary guidance suggests that early clinical intervention, at least to the extent of obtaining a diagnosis and preparing for future contingencies, is generally beneficial [21,45].

In the early stages of recognition, accepting "home help" from someone outside the family is a significant challenge faced by persons with dementia, family caregivers, and the homecare workers themselves. Home support from external agencies can take the form of medical services, household care, personal care, and respite care [46]. The transition into paid help has been associated with an increase in worry and strain for caregivers, but its sustained use has also been associated with reduced feelings of overload. Clare [47] reported participants' awareness of trying to find a balance between accepting help from others and becoming too dependent on it. The importance of informal support and other social networks for people with dementia and their caregivers has been studied by Carpentier [48,49], using social network analysis.

Another significant point of transition for families living with dementia is the reduction and eventual cessation of driving. For people with dementia, this is particularly challenging because of impaired insight, judgment, and reasoning skills [50] mixed with visual perceptual difficulties. Up to 25% of older adults continue to drive after being advised by a physician to stop [51]. While drivers with dementia are at increased risk for road traffic accidents [52], there has been a relative lack of research on the psychosocial issues associated with driving cessation and on factors that could mediate the impact of transportation dependence on subjective well-being [53].

Subjective well-being is also greatly affected by perceived autonomy over one's finances, both of which are affected in dementia because of the cognitive skills necessary to manage money. Most research examining the financial autonomy of older people with dementia has centered around concerns related to financial abuse [54,55] or issues related to capacity and power of attorney or guardianship [56–58]. Increasingly, there is interest in exploring how financial control can be at least partially maintained for as long as possible.

Admission to hospital is often a negative experience for people with dementia and their caregivers. Acute hospitalization of people with dementia has been studied in relation to avoiding transfer to hospital from long-term care homes [59], models of care for persons with dementia, making hospitals more dementia friendly, and access to rehabilitation.

Most people want to live in their own homes for as long as possible. When people with dementia are no longer able to manage at home independently, they may move to live with family, to an assisted living residence or to a long-term care home. The process of moving to a long-term care home has received more attention in research [60–62].

Problematic issues in the quality of end-of-life and palliative care for people with dementia include inadequate pain management [63], inappropriately aggressive medical care [64,65], the decision to hospitalize [59], the use of feeding tubes for nutrition and hydration [66,67], and the use of advance directives to guide decision-making [68,69]. Some have argued that palliative approaches might be beneficial early in disease progression so as to allow all those involved to focus on maintaining or improving quality of life rather than emphasizing health status itself.

Thus while most research has been undertaken into specific dementia-related transitions *in isolation*, some has sought to investigate linkage between transitions in terms of correlations between types of transition, patterns of effective transitional coping behavior, patient and caregiver experiences, and service provision and outcomes, identifying problems with health system effectiveness and efficiency. These challenges are common around the world, and much may be learned about efficient and effective service delivery by comparing structures, processes, and outcomes across countries with similar demographics and health systems [70,71], considering gaps in knowledge associated with transitions related to dementia and exploring the attendant needs of patients, caregivers, health-care providers, and policymakers.

Understanding determinants of transition quality for people with dementia

This book is fundamentally about successfully navigating through common and significant transitions, changes, and developments which are encountered by most, if not all, people with dementia and their caregivers. To understand this better, several of the authors contributing to this book undertook a study called the Dementia Transitions Study designed to answer the following question: Through the course of the disease, are clinical and social variables associated with the perceived quality of transitions experienced by people with dementia and their caregivers? [72].

The study integrated a prospective longitudinal design with a cross-sectional aspect, to explore transitions as they occurred. People with dementia and their principal caregivers were recruited as "dyads." In-person contacts were arranged with participants every 6 months for up to 24 months to obtain repeated-measures data and assess whether any of the transitions of interest had occurred. The study measures allowed for caregivers to report on the condition of those with dementia. Caregivers also reported on their own condition.

In addition, every month each dyad was contacted by telephone. A combination of a structured questionnaire followed by semistructured interview with open-ended questions was used to elicit information about any significant changes in status or circumstance with respect to the index transitions. These data were used to assess transition quality. Further qualitative analyses of some of these data are incorporated in several chapters of this book.

To identify the "quality" of the process associated with each of the transitions, an adjudication panel independently reviewed the interview transcripts. In making these assessments, the judges considered a general question about whether the quality of each transition had been essentially "good," "bad," or whether they were "unable to decide." This question was answered from the perspective of the caregiver. It was also considered from the perspectives of the judges, taking into account the interview data, their knowledge of the disease and of the Canadian health system. This was designated as the "summative" transition quality outcome variable.

In all, 108 dyads were enrolled in the study. Of these, 90 experienced a total of 166 transitions during the 2-year study period (mean = 0.92 transitions/dyad/year), of whom 87 people with dementia experienced 157 transitions which could be judged as being "good" or "bad" by the investigators. Nine transitions either lacked data or the assessors were unable to make a judgment. Forty-one people with dementia experienced only one type of transition during the study period. The other 46 experienced between two and five types of transition.

Drummond et al. [72] reported that of the 157 transitions with a summative outcome, 107 (68.2%) were judged as being "good." Of the 149 transitions with caregiver-reported data, 111 transitions (74.5%) were judged to be "good." Most transition types had more experiences of "good"-quality transitions than bad. The exception to this was "acute hospital admission," which was associated with more negative than positive outcomes.

Caregivers' assessments were generally less critical about the quality of the transition than the summative ones were. This may reflect differences in expectations of care: Caregivers' expectations may be low, but they may see themselves as obtaining some benefit; professional providers may expect a certain performance standard and have no personal stake in the experience. Dementia is an important risk factor for patients being assigned to alternate level of care beds in hospital, staying longer than needed while waiting for an appropriate place to be discharged to [73]. Poor admission experiences suggest gaps relating both to the provision of care [74–76] and to patient and caregiver engagement in policy developed for it, with implications for quality of life, activities of daily living, and caregiver burden. They imply that pragmatic screening for dementia should be undertaken at the very beginning of the admission process for older patients and that the approach to managing admission should recognize and respond to the vulnerability of those who are cognitively challenged. In practice, this might involve simply asking caregivers whether the person being admitted has symptoms of dementia and if so allowing more time for admission processes. Identifying caregivers (and patients whenever feasible) as members of the care-providing team may be a crucial resource if it enables prevention of acute hospital admission. Preventing admission in people with dementia should be a priority in its own right. See Chapter 5 for more discussion of problems and solutions regarding persons with dementia's experiences of inpatient hospital treatment.

Drummond et al. [72] also found evidence that better baseline competence in activities of daily living and quality of life in people with dementia was associated with better perceived transition quality. Higher caregiver burden was associated with perceived poorer transition quality as reported by caregivers. People with dementia who have worse prior experience on these dimensions require careful attention, particularly in primary care settings and in preparing and supporting them before, during, and after transitions throughout the course of the disease.

The distinction between cognitive determinants (which were not associated with transition quality) and functional determinants (which were) reinforces understanding that functional challenges are more easily recognized in people with dementia than cognitive ones and that functional capacity has a stronger self-reported influence on "life satisfaction" in seniors than cognition [77]. Hence, providers should pay close attention to activities of daily living as being predictive of caregiver burden and poorer patient experience in general. Training and support for managing dementia-related transitions is important for professionals and caregivers alike. Recent work exploring contemporary research priorities among patients and caregivers supports an integrated, team-based, person-oriented approach [78].

What does "good" mean in the context of transitions and dementia

Until recently, messaging about dementia has been mostly negative. We had focused on the fact that there is no cure, that there is only decline, and that people with dementia can only expect an increased number of horrible experiences, leaving them a burden to caregivers and society at large. This messaging has been successful in pushing societies, policymakers, and researchers to find solutions to prevent, care for, and live with dementia. But given this messaging, it is not surprising that our expectations of our experience with dementia will be shaped by this bias. At the core of this book are ideas, practices, and structural contexts about what it takes to have and maintain good quality of life despite the onset and progression of dementia. Expectations, as well as hope, trust, optimism, and resilience, are all important qualities that may change the trajectories of dementia experience or how we 'ride the waves'. All are multidimensional. They are also interrelated, rather than independent constructs. It may be helpful to think of them as dimensions of the larger quality of life construct. For example, one might think about quality of life as expressed by resilience, while at the same time considering resilience as the expression of the multiple interrelationships between expectation, trust, optimism, hope, and other dimensions of quality of life, just as one might consider hope as the expression of the multiple interrelationships between resilience, expectation, optimism, and so on.

Hope

Hope may be understood as a preference with at least some expectation of future fulfillment. Hope is an important aspect of living with chronic conditions, for example Amyotrophic Lateral Sclerosis (ALS) [79] and cancer [80]. Information about how to stay positive has only recently become part of public messaging about dementia. Yet it is fundamental to how people with dementia and those who surround them navigate transitions.

The influence of hope in lived experiences of dementia has been studied by Wolverson et al. [81], who identified seven themes which define hope in the social science literature. These include:

- a positive future orientation
- a sense of desire or preference
- hoping for a specific outcome
- generalized hope in the future
- hoping for a better future than current circumstances provide
- a sense of power or motivation to create an improved future, and
- hope deriving from the social support of other people.

They interviewed 10 people with early-stage dementia about their experiences of hope in the context of their health condition. These broadly fell into two overarching themes. The first was a process of maintaining hopefulness, likely influenced by family history and upbringing, being realistic about their difficulties associated with dementia and adjusting their expectations accordingly. The second theme included the influence of a hopeful community or society in general. "Keep living" indicated motivation to maintain a preferred quality of life, largely through ensuring that valued relationships and social connections continued and that day-to-day life remained a positive experience.

There is a sense among providers that even if dementia-related cognitive decline is irreversible, there are other aspects of living which may be improved in people with the disease and which, therefore, constitute grounds for recovery and evidence for well-being [82]. The "recovery movement" in mental illness treatment [83] sees it as a journey and process connected to meaning in life, self-identity, relationships, partnership, empowerment, engagement, and allowing and supporting people to manage their own health according to their own preferences even when symptoms persist or recur [83–85], including those of dementia [80]. Implementing these principles into everyday practice depends on the motivation and commitment of all involved, and hopefulness is central to the process.

Other research has identified links between hope and aspects of quality of life of people with dementia. In a systematic review, Agli et al. [86] note that hopefulness associated with religious and spiritual sensibility are positive resources, which protect and maintain the quality of life of people with dementia. Higher self- or caregiver-reported levels of hope were associated with lower levels of anxiety among residents in long-term care homes, including among people with dementia [87].

Spector and Orrell [88] explored quality of life in people with dementia by comparing the self-ratings of patients with the proxy ratings by staff in nine long-term care facilities in the UK. They used the "hope" subscore of the Approaches to Dementia Questionnaire [89]. There was a statistically significant association between staff hope and patient-reported quality of life. Sites with higher mean staff hope scores had higher mean patient-reported quality of life scores. This association does not imply that hopeful staff cause better patient quality of life, although it is possible. More hopeful staff may engage more fully and constructively with residents. The nature of the association requires further research, but the evidence does suggest that hopefulness in staff and quality of life of patients are linked, and not by chance.

Hope is also important for quality of life of caregivers. Duggleby et al. [90] found that caregivers living at home with persons with dementia experienced hope fading and they worked to replenish it, for example, by identifying positive aspects of their experience, by connecting with others, through prayer, and by being optimistic. Hope was described as an important resource to fuel continued caregiving.

In a mixed-method study of hope and quality of life of caregivers during transitions, Duggleby et al. [91] found an interaction between coping and hope. Caregivers with higher hope scores who sought information and assistance during transitions had higher quality of life. Higher psychological quality of life was associated with higher scores on two of three dimensions on the Herth Hope Index [92]: the 'interconnectedness' dimension, including faith, inner strength and reciprocal caring relationships—and the 'temporality and future' dimension, including having positive outlook and future goals.

As will be seen in subsequent chapters, for several transitions, hope does seem to play a role in the transition experiences of persons with dementia and caregivers. For example, hope influences decisions to stay at home in the face of decreasing ability to manage at home [93] and hopefulness makes it easier to accept the decision to move. Hopefulness (and optimism) seems to be associated with better adjustment after the move to an assisted living residence [94]. A hope-focused group for residents of long-term care homes was found to be feasible and acceptable, with indications that it may have renewed and sustained hope for participants [95]. Hope-focused interventions may be a way to provide support for people with dementia through transitions involving loss, such as loss of driving, moving to an assisted living residence, or moving to a long-term care home.

Expectancy

Hope is about the desirability of, or preference for, a future state, whereas expectation is about the probability of a state being actualized. The more certain a future outcome is judged to be, even subjectively, the stronger the expectation that it will occur. Leung et al. [96] developed a conceptual model for understanding how these two approaches to futures (i.e., hopes and expectations) relate to each other. In brief, hopes and expectations are congruent when the probability of a hoped for outcome is high. People are motivated to increase chances of achieving hoped for outcomes, thus maintaining congruence between hope and expectation. Many factors influence the extent to which hopes and expectations are congruent, including aspects of the hoped for outcome and factors internal and external to the person. When an expectation is hoped for and fulfilled, the individual experiencing it perceives benefit. When an expectation is not hoped for but fulfilled, the individual experiences no benefit beyond the sense of correctly predicting the outcome and may experience disappointment to varying degrees, depending on the degree of preference attached to the expectation.

Hopes and expectations are often confused as being the same thing. They are both assessments of future experience, but they express different ways of thinking about the future—preference and probability, respectively. Thus it is useful to think of them as separate and independent constructs, influenced by similar kinds of factors but in different ways. For example, public messaging about decline associated with dementia may create expectations about the need to seek medical care and also influence unrealistic expectations about and diminished hope of living well with the condition.

"Health expectations" are a particularly important example of the relationship between hope and expectations. The ways that they influence our behavior are well illustrated in the literature on prescribed medication. Understanding that patient expectation has a significant influence over physician prescribing behaviors began to emerge in the 1960s. Scheff [97] adopted the concept in his investigation of diagnostic error, remarking that both doctors and patients were "biased toward the expectation of treatment." Research in the 1990s provided evidence that patient expectations as well as physicians' perceptions of patients' expectations of treatment influence prescribing behavior [98–103]. Patients with an expectation that they would receive medication were three times more likely to receive a prescription and they were 10 times more likely to receive a prescription if their physician believed that they expected medication. Recent research confirms that physicians' beliefs about patient or caregiver expectation influence their willingness to deprescribe [104], particularly if patients are believed to expect that medications will continue to provide future benefit [105]. Thus how people expect their dementia trajectories to unfold, irrespective of how preferable those trajectories may be, may influence how providers behave. Knowing what we expect may, therefore, generate a degree of control over the health care we receive.

The same can be said about how our health expectations can influence our own behaviors. Placebo is typically defined as "any therapy or component of therapy used for its nonspecific, psychological, or psychophysiological effect, or that is used for its presumed specific effect, but is without specific activity for the condition being treated" [106]. Despite previous evidence that placebo effects are substantial (around 35%) [107], more recent research [108] indicates that although they are statistically significant (i.e., unlikely to be due to chance), their magnitude may not be large.

Evidence for placebo effects in dementia is best found in randomized controlled trials of medications. Given the understanding that placebo effects are common, the evaluation of a medicine's clinical effectiveness has to take account of placebo effects before the benefit of a medicine may be estimated. In a study examining the efficacy of donepezil compared with placebo in patients with mild vascular dementia, a statistically significant difference in cognition in favor of the medication was measured at 12, 18, and 24 weeks, but much of the difference derived from a diminution in performance in the placebo group at 24 weeks; at 12 and 18 weeks, both those in the treatment and placebo groups reported improved cognition [109].

An interesting aspect of the placebo effect is evidence that it has increased over time. A systematic review and meta-analysis of placebo effects in clinical trials for medications used in the management of behavioral and psychological symptoms of dementia found that participants receiving placebo in recent studies reported greater benefits than those reported by participants in earlier studies [110]. The authors discuss a number of possible explanations for this increase in the size of placebo effects, one of which is a suggestion that expectation might generate stronger placebo effects, in that expectations of the effectiveness of medications might have increased in the community through time, leading to higher ratings of outcome by those taking placebo and those receiving the active drug.

These issues raise questions about the precise nature of expectations and placebo effects. Some evidence, particularly from research into pain relief and antidepressants, suggests that the placebo mechanism involves an interaction between patient and health-care provider expectancies about the probable effectiveness of the treatment and that expectancy may have effects on brain neurotransmitters [111,112]. Rutherford et al. [112] also suggest the possibility of indirect processes, whereby expectancy influences health behaviors, such as medication adherence, or improves the quality of the relationship between a patient and his/her health-care provider.

Rutherford et al. [112] found possible age-related differences in the relationship between expectancies about antidepressant medication and subsequent symptom experience. Among those aged 55–65 years, treatment expectancies were not associated with symptom experiences, whereas high expectancy was associated with better symptom improvement for younger patients. One interpretation the authors made for this finding is that vascular damage in the brain (as in dementia) may inhibit the influence of expectancy about depression symptoms, suggesting that further investigation, including with persons with dementia, may be worthwhile.

Some research has investigated expectation in association with transition experiences of people with dementia and their caregivers, primarily focusing on expectations about the transition or the health-care encounter. Interviews of patients and caregivers about their expectations on referral to a memory clinic revealed substantial differences between them [113]. Patient expectations varied depending on their understanding of the purpose of the clinic assessment. Some had vague expectations of receiving heath advice, some expected to be reassured that there was no problem with their memory, and others expected to find out the reason for their memory problems, get a diagnosis, and receive help. Similar to findings of Morgan et al. [114], caregivers expected to have their concerns about the person with dementia validated. They tended to be dissatisfied because their expectations about obtaining information, advice, and "tangible solutions" were not met. Similar mismatch between caregiver expectations and experiences of health-care services during transitions have been identified with hospitalization [115], admission to assisted living [116], and admission to long-term care homes (See Chapters 5, 7, and 8).

Read et al. [117] interviewed people with dementia about their expectations for the future. Uncertainty about what to expect as dementia progresses contributed to feelings of anxiety and worry. This uncertainty meant that they were uncertain about what their care needs would be, making it difficult to engage in advanced care planning. This might have an effect on their readiness for and acceptance of moving to a long-term care home or assisted living residence.

Uncertainty about what will happen as dementia progresses has also been identified as an issue for caregivers of people with dementia in long-term care homes [118]. Staff report that some caregivers expect that the person with dementia will improve and do not recognize that dementia is a terminal illness, making end-of-life planning discussions difficult. Many caregivers want information about what to expect at end of life and find it difficult to obtain this information [119].

If we accept that expectancies among people with dementia and their caregivers are important contributors to their mutual experience of quality of life, then it makes sense that recognition of these expectations and discussions about them should become an active part of decision-making and clinical discussions between people with dementia, caregivers, and providers.

Trust

Expectancy, preference and trust, another major contributor to quality of life, are conceptually linked. Trust is a quality deriving from familiarity, mutuality, predictive certainty, shared knowledge and experience. Lee [120] identifies two broad definitions of trust. Both concern aspects of the quality of relationships which people form with other people as individuals or groups, or with other "entities." The first definition conceptualizes trust as a degree of perceived certainty that future states will be actualized. These future states may be benign or beneficial, or they may involve hazard. Hence statements such as "This person will probably help me," "That person will certainly harm me," "This bridge might be safe to cross," and

"Oh no it won't," all express degrees of trust in future states. But commonly, trust tends to be expressed in relation to positive future states deriving from people acting in a beneficial manner. We typically express trust in positive futures, not in negative ones. We say "I trust you to help me," not "I trust you not to help me." In this conceptualization, trust expresses aspects of expectancy *and* preference. Lee [120] identifies 22 components of trust associated with facilitating public participation in social institutions, and each reflects a moral "good" or value, including "transparency," "accessibility," "equitability," "safety," "responsiveness," and "accountability."

In thinking about quality of life in people with dementia and their caregivers, one major social institution for which a trusted relationship is clearly important is individuals and organizations providing health care. Older immigrants to Canada often experience serious difficulty accessing health care through a combination of language problems and sociocultural norms which militate against the establishment of trusting relationships [18]. Koehn et al. [121] used the candidacy framework of Dixon-Woods [122] to study the role of trust in achieving access to care for people with dementia in Korean and Punjabi ethnolinguistic communities in Vancouver, Canada. They concluded that although community-based agencies, particularly those from the multicultural and settlement sector, are more often instrumental in establishing trusted relationships between their staff and clients, they often lack detailed knowledge about heath conditions and their treatment and management and they lack power to implement statutory care. Hence partnerships between mainstream mental health/dementia services and the community sector have proven successful in increasing the accessibility of specialized resources while maximizing their combined trustworthiness, accessibility, and effectiveness in association with greater certainty that preferences will be realized.

Optimism

Other related constructs include psychological components of quality of life associated with "fulfilment." If we ask the rhetorical question "What do we need to live a good life?" many people will say that remaining positive in the face of challenging circumstances or conditions is key. Optimism is a trait which all people, except the truly despairing, exhibit to some extent, although some more than others. Defined broadly as a general conviction of being able to exercise control when circumstances allow it, and to achieve success when doing so [123], optimism has been described as a "generalized expectancy of control" [124]. One might argue that optimism is more basically a generalized expectancy of utility, whether derived from controlled actions or not. For instance, if somebody keeps our glass half full without us doing anything to cause them to do it, we are probably going to become increasingly optimistic about future glass-fillings, even if we do not think we control them. But many, and perhaps most people, do regard control and agency as "goods," so a definition of optimism which includes those concepts is useful. Optimists may be described as people in whom high expectancies of preferred outcomes are more frequent than low expectancies of preferred outcomes.

Ruisoto et al. [123] studied optimism and quality of life in caregivers of people with dementia and found that quality of life (defined using the World Health Organization Quality of Life Scale-Brief instrument [125] was positively correlated with optimism and negatively correlated with caregiver burden. Optimism consistently predicted four dimensions of health-related quality of life (physical, psychological, social, and environmental) even after adjusting for sociodemographic, clinical (in the care receiver), and burden variables. In a Brazilian study, caregivers with "grounded optimism" were less likely to use dementia day programs and more likely to engage in care at home [126]. Ploeg et al. [127] reported that optimism is important for caregivers during care-related transitions.

Self-efficacy

Optimism was defined by Ruisoto et al. [123] to include elements of control. Self-efficacy explains the idea that we are more likely to adopt a behavior if we believe we are able to execute it [128]. Many studies have used this concept to understand associations and causal relationships between behaviors and outcomes. There are several well-validated scales, both generic and condition-specific, intended to estimate strength or magnitude of self-efficacy in individuals. In relation to dementia, Fortinsky et al. [129] developed a set of 10 questions designed specifically to elicit information on caregiver self-efficacy. Nine of the self-efficacy items clustered into two factors concerned with symptom management and use of "community service support." Caregivers with higher symptom management self-efficacy scores reported fewer depressive symptoms, and caregivers with higher service use self-efficacy scores reported fewer physical health problems of their own.

Since publication of the Fortinsky paper, substantial research has been carried out into the determinants of self-efficacy associated with dementia caregiving. In research about the influence of self-efficacy on distress experienced by caregivers of people experiencing behavioral and psychological symptoms of dementia, Nogales–Gonzales et al. [130] showed that higher caregiver self-efficacy was associated with attenuation of the impact of behavioral problems on caregiver distress. Hence self-efficacy is an important attribute of quality of life in caregivers of people with dementia, and improving self-efficacy is a goal of many caregiver support interventions. So far, very little research has been undertaken into the impact of self-efficacy among people with dementia themselves. It may be challenging to conduct this research with people in the later stages of the disease, but it is important to study people in the earlier stages who have insight into their condition and have expectancies and preferences for how they would like to live. This is a gap in our knowledge. Most of the existing studies are cross-sectional and descriptive, identifying non-chance associations between self-efficacy and outcomes but not able to indicate causality. We do not know whether self-efficacy causes the outcomes or the other way round, nor whether other factors confound associations that we do see.

Locus of control

Related to notions of self-efficacy is locus of control [131], also important to perceived quality of life. It is a psychological concept describing the extent to which individuals see themselves as having personal (or "internal") control over events in their lives or to which control lies in others ("external"). In health-related research, the concept is often used to understand the extent to which people think that their health status and future health outcomes are achievable by their own actions. Locus of control is now generally considered to be one of four psychological "core self-evaluation" concepts, along with neuroticism, self-efficacy, and self-esteem, through which individuals judge their worth as people and the general quality of their lives. They may also be considered as being among the predictors of resilience [132].

As with self-efficacy, studies of associations between dementia-related outcomes and locus of control have focused on caregivers rather than on people with dementia themselves. The results have been somewhat varied. Nordtug et al. [133] studied the influence of neuroticism and external locus of control in caregivers of people with dementia and found that while neuroticism was predictive of increased burden and mental health problems, external locus of control was not. But Bruvik et al. [134] reported that locus of control was the main predictor of burden in caregivers of people with dementia, with greater internalized control associated with lower burden and greater externalized control associated with greater burden.

Some support for loss of control as a perceived cause for behavioral and psychological symptoms of dementia is provided by Polenick et al. [135], who applied attribution theory in an analysis of focus group data. Caregivers attributed such symptoms in people with dementia to their fear at their expected loss of control.

Research about "control" in persons with dementia has frequently been less focused on the psychological concept of locus of control and more on perceived autonomy. As will be seen in subsequent chapters, the extent to which persons with dementia have control of, or input into, decisions about their lives varies. Autonomy and control influence transition experiences. Those with dementia often feel that decisions about their lives are made for them rather than with them [136].

Conclusion

These dimensions of quality, and their mutual interrelationships, provide insight into what we mean when we talk about good or bad transitions for people with dementia and shed light on methods, structures, and processes which are likely to maximize the former and minimize the latter. As noted earlier, conditions that promote strong probabilities of preferred states generate confidence in beneficial outcomes which individuals and groups have a sense are achievable. Collectively, these create good states. Lack of every one of those conditions probably leads to bad ones. In between are an infinite number of states which are more or less better or worse depending on the details and the people experiencing them. Receiving a diagnosis of dementia is challenging and difficult. But it is rarely a sentence of imminent, major cognitive decline and death. There is still much of life to be lived and enjoyed. Helping professionals providing care, care partners giving it, and people with dementia living through it themselves to each maximize its quality to the greatest extent possible is the fundamental goal of this book.

References

[1] Alzheimer Disease International. About Dementia. 2019. Available at: www.alz.co.uk/about-dementia.

[2] Parmar J, Dobbs B, McKay R, Kirwan C, Cooper T, Marin A, Gupta N. Diagnosis and management of dementia in primary care. Exploratory study. Can Fam Physician 2014;60(5):457–65. http://www.cfp.ca/.

[3] Chambers LW, Bancej C, McDowell I. Prevalence and monetary costs of dementia in Canada: population health expert panel. 2016. Retrieved from: http://www.alzheimer.ca/ab/~/media/Files/national/Statistics/Prevalence andCostsofDementia_EN.pdf.

[4] Drummond N, Birtwhistle R, Williamson T, Khan S, Garies S, Molnar F. Prevalence and management of dementia in primary care practices with electronic medical records: a report from the Canadian primary care sentinel surveillance network. CMAJ Open 2016;4(2):E177–84. https://doi.org/10.9778/cmajo.20150050.

[5] Aguerro Torres H, Fratiglione L, Guo Z, Viitanen M, Winblad B. Dementia in advanced age led to higher mortality rates and shortened life. J Clin Epidemiol 1999;52:737–43.

[6] Alzheimer's Research UK. Types of dementia. 2019. Available at: https:www.alzheimersresearchuk.org/about-dementia/types-of- dementia.

[7] Alonso A, et al. Cardiovascular risk factors and dementia mortality: 40 years of follow-up in the Seven Countries Study. J Neurol Sci 2009;280(1–2):79–83.

[8] Norton S, et al. Potential for primary prevention of Alzheimer's disease: an analysis of population-based data. Lancet Neurol 2014;(8):788.

[9] Rastas S, et al. Vascular risk factors and dementia in the general population aged >85 years: prospective population-based study. Neurobiol Aging 2010;31(1):1–7.

[10] Beydoun MA, et al. Association of adiposity status and changes in early to mid-adulthood with incidence of Alzheimer's disease. Am J Epidemiol 2008;168(10):1179–89.

[11] Richard E, et al. Methodological issues in a cluster-randomized trial to prevent dementia by intensive vascular care. J Nutr Health Aging 2010;14(4):315–7.

[12] Viswanathan A, Rocca WA, Tzourio C. Vascular risk factors and dementia: how to move forward? Neurology 2009;72(4):368–74.

[13] Fillit H, et al. Cardiovascular risk factors and dementia. Am J Geriatr Pharmacother 2008;6:100–18.

[14] Barnes DE, Yaffe K. The projected effect of risk factor reduction on Alzheimer's disease prevalence. Lancet Neurol 2011;(9):819.

[15] Brookmeyer R, Johnson E, Ziegler-Graham K, Arrighic M. Forecasting the global burden of Alzheimer's disease. Alzheimer's Dementia 2007;3:186–91.

[16] Sposato LA, et al. Declining incidence of stroke and dementia: coincidence or prevention opportunity? JAMA Neurol 2015;72(12):1529.

[17] Matthews FE, et al. A two-decade comparison of prevalence of dementia in individuals aged 65 years and older from three geographical areas of England: results of the cognitive function and ageing study I and II. The Lancet 2013;(9902):1405.

[18] Koehn S, Badger M, Cohen C, McCleary L, Drummond N. Negotiating access to a diagnosis of dementia: implications for policies in health and social care. Dementia 2016;15(6):1436–56. https://doi.org/10.1186/1471-2296-11-52.

[19] Alzheimer Society of Ireland. Stages and Progression. 2019. Available at: https://www.alzheimer.ie/About-Dementia/Stages-progression.aspx.

[20] Statistics Canada. Deaths and causes of death. 2017. Available at: https://www150.statcan.gc.ca/n1/daily-quotidien/181129/dq181129g- cansim-eng.htm.

[21] Gauthier S, Patterson C, Chertkow H, et al. Recommendations of the 4th Canadian consensus conference on the diagnosis and treatment of dementia (CCCDTD4). Can Geriatr J 2012;15(4):120–6. https://doi.org/10.5770/cgj.15.49.
[22] Lanctot KL, Hermann N, Yau KK, et al. Efficacy and safety of cholinesterase inhibitors in Alzheimer's disease: a meta-analysis. CMAJ (Can Med Assoc J) 2003;169:557–64.
[23] Howard R, McShane R, Lindesay J, Ritchie C, Baldwin A, Barber R, Burns A, Dening T, Findlay D, Holmes C, Hughes A, Jacoby R, Jones R, Jones R, McKeith I, Macharouthu A, O'Brien J, Passmore P, Sheehan B, Juszczak E, Katona C, Hills R, Knapp M, Ballard C, Brown R, Banerjee S, Onions C, Griffin M, Adams J, Gray R, Johnson T, Bentham P, Phillips P. Donepezil and memantine for moderate-to-severe Alzheimer's disease. N Engl J Med March 2012;366(10):893–903.
[24] Braun M, Scholz U, Bailey B, Perren S, Hornung R, Martin M. Dementia caregiving in spousal relationships: a dyadic perspective. Aging Ment Health 2009;13(3):426–36. https://doi.org/10.1080/13607860902879441.
[25] Kaplan DB, Andersen TC, Lehning AJ, Perry TE. Aging in place vs. Relocation for older adults with neurocognitive disorder: applications of Wiseman's behavioral model. J Gerontol Soc Work 2015;58(5):521–38. https://doi.org/10.1080/01634372.2015.1052175.
[26] Ferri CP, Prince M, Brayne C, Brodaty H, Fratiglioni L, Ganguli M, et al. Global prevalence of dementia: a Delphi consensus study. Lancet 2005;366:2112–7.
[27] Canadian Study of Health and Aging Working Group. Canadian study of health and aging: study methods and prevalence of dementia. CMAJ (Can Med Assoc J) 1994;150(6):906–12.
[28] American Psychiatric Association. In: Diagnostic and statistical manual of mental disorders. 4th ed., text revision. Washington, DC: Am Psychiatric Assoc; 2000.
[29] Callahan CM, Boustani MA, Unverzagt FW, Austrom MG, Damush TM, Perkins AJ, et al. Effectiveness of collaborative care for older adults with Alzheimer disease in primary care: a randomized controlled trial. J Am Med Assoc 2006;29(18):2148–57.
[30] Prigerson HG. Costs to society of family caregiving for patients with end- stage Alzheimer's disease. N Engl J Med 2003;349(20):1891–2.
[31] Ylieff M, Buntinx F, Fontaine O, De Lepeleire J. Long-term assistance and care for dependent elderly and people with dementia. Arch Public Health 2004;62:117–24.
[32] Coleman EA, Boult C. Improving the quality of transitional care for persons with complex care needs. J Am Geriatr Soc 2003;51:556–7.
[33] Dementia. In: Merriam-Webster; 2019. Available at: www.merriam-webster.com.
[34] Bridges W. Transitions: making sense of life's changes. USA: Addison-Wesley Publishing Co., Inc; 1980.
[35] Meleis AI, Sawyer LM, Im E, Messias DKH, Schumacher K. Experiencing transitions: an emerging middle-range theory. Adv Nurs Sci 2000;23(1):12–28.
[36] Schumacher KL, Meleis AI. Transitions: a central concept in nursing. Image J Nurs Scholarsh 1994;26(2):119–27.
[37] Schumacher KL, Jones PS, Meleis AI. Helping elderly persons in transition: a framework for research and practice. In: Swanson E, Tripp-Reimer T, editors. Life transitions in the older adult: issues for nurses and other health professionals. New York: Springer Publishing Company; 1999. p. 1–25.
[38] Adams KB. Transition to caregiving: the experience of family members embarking on the dementia caregiving career. J Gerontol Soc Work 2006;47(3–4):3–29.
[39] Clare L. Developing awareness about awareness in early-stage dementia. The role of psychosocial factors. Dementia 2002a;1(3):295–312.

[40] Clare L. The construction of awareness in early-stage Alzheimer's disease: a review of concepts and models. Br J Clin Psychol 2004;43:155–75.
[41] De Lepeleire J, Heyman J, Buntinx F. The early diagnosis of dementia: triggers, early signs and luxating events. Fam Pract 1998;15(5):431–6.
[42] De Lepeleire J, Heyman J. Diagnosis and management of dementia in primary care at an early stage: the need for a new concept and an adapted procedure. Theor Med Bioeth 1999;20:215–28.
[43] Phinney A. Fluctuating awareness and the breakdown of the illness narrative in dementia. Dementia 2002;1(3):329–44.
[44] Phinney A, Wallhagen M, Sands LP. Exploring the meaning of symptom awareness and unawareness in dementia. J Neurosci Nurs 2002;34(2):79–90.
[45] Iliffe S, Manthorpe J, Eden A. Sooner or later? Issues in the early diagnosis of dementia in general practice: a qualitative study. Fam Pract 2003;20(4):376–81.
[46] Pot AM, Zarit SH, Twisk JWR, Townsend AL. Transitions in caregivers' use of paid home help: associations with stress appraisals and well-being. Psychol Aging 2005;20(2):211–9.
[47] Clare L. We'll fight it as long as we can: coping with the onset of Alzheimer's disease. Aging Ment Health 2002b;6(2):139–48.
[48] Carpentier N, Ducharme F. Care-giver network transformations: the need for an integrated perspective. Ageing Soc 2003;23:507–25.
[49] Carpentier N, Ducharme F. Support network transformations in the first stages of the caregiver's career. Qual Health Res 2005;15(3):289–311.
[50] Adler G, Kuskowski M. Driving habits and cessation in older men with dementia. Alzheimers Dis Assoc Disord 2003;17:68–71.
[51] Dobbs B, Carr DB, Morris JC. Management and assessment of the demented driver. The Neurologist 2002;8:61–70.
[52] Fox GK, Bowden SC, Bashford GM, Smith DS. Alzheimer's disease and driving: prediction and assessment of driving performance. J Am Geriatr Soc 1997;45:949–53.
[53] Dickerson AE, Molnar LJ, Eby DW, Adler G, Bedard M, Berg-Weger M, Classen S, Foley D, Horowitz A, Kerschner H, Page O, Silverstein NM, Staplin L, Trujillo L. Transportation and aging: a research agenda for advancing safe mobility. Gerontol 2007;47(5):578–90.
[54] Heath JM, Brown M, Kobylarz RA, Castano S. The prevalence of undiagnsoed geriatric health conditions among adult protective service clients. Gerontol 2005;45(6):820–3.
[55] Rabiner DJ, Brown D, O'Keefe J. Financial exploitation of older persons: POlicy issues and recommendations for addressing them. J Elder Abuse Negl 2004;16(1):65–83.
[56] Jones RG. The law and dementia—issues in England and Wales. Aging Ment Health 2001;5(4):329–34.
[57] Moye J, Marson DC. Assessment of decision-making capacity in older adults: an emerging area of practice and research. J Gerontol B 2007;62B(1):P3–11.
[58] Brechling BG, Schneider CA. Preserving autonomy in early stage dementia. J Gerontol Soc Work 1993;20(1–2):17–33.
[59] Mitchell SL, Teno JM, Intrator O, Feng Z, Mor V. Decisions to forgo hospitalization in advanced dementia: a nationwide study. J Am Geriatr Soc 2007;55:432–8.
[60] Liken MA. Managing transitions and placement of caring for a relative with Alzheimer's Disease. Home Health Care Manag Pract 2001;14(1):31–9.
[61] Mead LC, Eckert JK, Zimmerman S, Schumacher JG. Sociocultural aspects of transitions from assisted living for residents with dementia. Gerontol 2005;45(1):115–23.

[62] Neumann PJ, Araki SS, Arcelus A, Longo A, Papadopoulos G, Kosik KS, et al. Measuring Alzheimer's disease progression with transition probabilities. Neurology 2001;57:957–64.
[63] Blasi ZV, Hurley AC, Volicer L. End-of-life care in dementia: a review of problems, prospects, and solutions in practice. J Am Med Dir Assoc 2002;3:57–65.
[64] Richardson SS, Sullivan G, Hill A, Yu W. Use of aggressive medical treatments near the end of life: differences between patients with and without dementia. Health Res Educ Trust 2007;42(1):183–200.
[65] Richter J, Eisemann MR, Zgonnikova E. Doctors' attitudes and end-of-life decisions in the elderly: a comparative study between Sweden, Germany and Russia. Arch Gerontol Geriatr 2002;34:107–15.
[66] Buiting HM, van Delden JJM, Riejens JAC, Onwuteaka-Philipsen BD, Bilsen J, Fischer S, Lofmark R, Miccinesi G, Norup M, van der Heide A. Forgoing artificial nutrition or hydration in patients nearing death in six European countries. J Pain Symptom Manag 2007;34(3):305–14.
[67] Cervo FA, Bryan L, Farber S. To PEG or not to PEG: a review of evidence for placing feeding tubes in advanced dementia and the decision-making process. Geriatrics 2006;61(6):30–5.
[68] Dresser R. Dworkin on dementia: elegant theory, questionable practice. Hastings Cent Rep 1995;22(6):32–8.
[69] Harvey M. Advance directives and the severely demented. J Med Philos 2006;31:47–64.
[70] Bond J, Stave C, Sganga A, O'Connell B, Stanley RL. Inequalities in dementia care across europe: key findings of the facing dementia survey. Int J Clin Pract 2005;146(Suppl. l):8–14.
[71] Drummond N. Enlarging the community of care for people with dementia: the potential of international collaborative research. Dementia 2007;6(2):171.
[72] Drummond N, McCleary L, Garcia L, McGilton K, Molnar F, Dalziel W, Xu TJ, Turner D, Triscott J, Freiheit E. Assessing determinants of perceived qualityin transitions for people with dementia: a prospectiveobservational study. Can Geriatr J 2019;22:13–22. https://doi.org/10.5770/cgj.22.332.
[73] McLoskey R, Jarrett P, Stewart C, et al. Alternate level of care patients in hospitals: what does dementia have to do with this? Can Geriatr J 2014;17:88–94.
[74] Timmons S, O'Shea E, O'Neill D, et al. Acute hospital dementia care: results from a national audit. BMC Geriatr 2016;16(1):113.
[75] Spencer K, Foster P, Whittamore KH, et al. Delivering dementia care differently—evaluating the differences and similarities between a specialist medical and mental health unit and standard acute care wards: a qualitative study of family carers' perceptions of quality of care. BMJ Open 2013;3(12):e004198.
[76] George J, Long S, Vincent C. How can we keep patients with dementia safe in our acute hospitals? A review of challenges and solutions. J Royal Soc Med 2013;106(9):355–61.
[77] St John PD, Montgomery PR. Cognitive impairment and life satisfaction in older adults. Int J Geriatr Psychiatry 2010;25(8):814–21.
[78] McGilton K, Bethell J. Announcing the top 10 Canadian dementia research priorities. Alzheimer Society Blog; 2017. Available at: http://alzheimersocietyblog.ca/top-10-research-priorities/.
[79] Cross KL. The biochemistry of hope. J Palliat Med 2011;14:982–3.
[80] Von Roenn JH, von Gunten CF. Setting goals to maintain hope. J Clin Oncol 2003;21:570–4.

[81] Wolverson EL, Clarke C, Moniz-Cook E. Remaining hopeful in early-stage dementia: a qualitative study. Aging Ment Health 2010;14:450–60. https://doi.org/10.1080/13607860903483110.
[82] Kitwood T, Bredin K. Towards a theory of dementia care: personhood and well-being. Ageing Soc 1992;12:269–87.
[83] Anthony WA. Recovery from mental illness: the guiding vision of the mental health service system in the 1990s. Psychosoc Rehabil J 1993;16:11.
[84] Leamy M, Bird V, Le Boutillier C, Williams J, Slade M. Conceptual framework for personal recovery in mental health: systematic review and narrative synthesis. Br J Psychiatry 2011;199:445–52.
[85] Onken SJ, Craig CM, Ridgway P, Ralph RO, Cook JA. An analysis of the definitions and elements of recovery: a review of the literature. Psychiatr Rehabil J 2007;31:9.
[86] Agli O, Bailly N, Ferrand C. Spirituality and religion in older adults with dementia: a systematic review 2015;27:715–25.
[87] Creighton AS, Davison TE, Kissane DW. The correlates of anxiety among older adults in nursing homes and other residential aged care facilities: a systemayic reveiw. Int J Geriatr Psychiatry 2017;32:141–54.
[88] Spector A, Orrell M. Quality of life (QoL) in dementia: a comparison of the perceptions odf people with dementia and care staff in residential homes. Alzheimers Dis Assoc Disord 2006;20:160–5.
[89] Lintern T, Woods B. Approaches to dementia questionnaire. Bangor, UK: University of Wales; 1996.
[90] Duggleby W, Williams A, Wright K, Bollinger S. Renewing everyday hope: the hope experience of family caregivers of persons with dementia. Issues Ment Health Nurs 2009;30(8):514–21.
[91] Duggleby W, Swindle J, Peacock S, Ghosh S. A mixed methods study of hope, transitions, and quality of life in family caregivers of persons with Alzheimer's disease. BMC Geriatr 2011;11:88.
[92] Haugen G, Utvaer BK, Moksnes UK. The Herth Hope Index: a psychometric study among cognitively intact nursing home residents. J Nurs Meas 2013;21:378–400.
[93] Thein NW, D'Souza G, Sheehan B. Expectations and experience of moving to a care home: perceptions of older people with dementia. Dementia 2011;10:7–18.
[94] Aminzadeh F, Dalziel W, Molnar F, Garcia L. Symbolic meaning of relocation to a residential care facility for persons with dementia. Aging Ment Health 2009;13:487–96. https://doi.org/10.1080/13607860802607314.
[95] Moore SL, Hall SE, Jackson J. Exploring the experience of nursing home residents participation in a hope-focused group. Nurs Res Pract 2014:1–9. Article ID 623082.
[96] Leung KK, Silvius JL, Pimlott N, Dalziel W, Drummond N. Why health expectations and hopes are different the development of a conceptual model. Health Expect 2009;12:347–60.
[97] Scheff TJ. Decision rules, types of error, and their consequences in medical diagnosis. Behav Sci 1963;8:97–107.
[98] Vinson DC, Lutz LJ. The effect of parental expectations on treatment of children with a cough: a report from ASPN. J Fam Pract 1993;37:23–7.
[99] Britten N. Patients' demands for prescriptions for drugs in primary care. BMJ 1995;310:1084–5.
[100] Cockburn J, Pit S. Prescribing behaviour in clinical practice: patients expectations and doctors' perceptions of patients' expectations—a questionnaire study. BMJ 1997;315:520–3.

[101] MacFarlane J, Holmes W, MacFarlane R, Britten N. Influence of patients' expectations on antibiotic management of acute lower respiratory tract illness in general practice: questionnaire study. BMJ 1997;315:1211.
[102] Butler CC, Rollnick S, Pill R, Maggs-Rapport F, Stott N. Understanding the culture of prescribing: qualitative study of general practitioners' and patients' perceptions of antibiotics for sore throats. BMJ 1998;317:637–42.
[103] Himmel W, Lippert-Urbanke E, Kochen MM. Are patients more satisfied when the receive a prescription? The effect of patient expectations in general practice. Scand J Prim Health Care 1997;15:118–22.
[104] Wallis KA, Andrews A, Henderson M. Swimming against the tide: primary care physicians' views on deprescribing in everyday practice. Ann Fam Med 2017;15:341–6.
[105] Luymes C, Boelhouwer N, Poortvliet RKE, de Ruijter W, Reis R, Numans ME. Understanding deprescribing of preventive cardiovascular medication: a Q-methodology study in patients. Patient Prefer Adherence 2017;11:975–84.
[106] Shapiro, A.K, Morris, L.A. The placebo effect in medical and psychological therapies, in: Garfield, S.L. Bergin, A.E (Eds.), Handbook of psychotherapy and behavior change (2nd ed.) pp369–410. Wiley: New York.
[107] Beecher HK. The powerful placebo. J Am Med Assoc 1955;159:1602–6.
[108] Hrobjartsson A, Gotzsche PC. Placebo interventions for all clinical conditions. Cochrane Database Syst Rev 2010;(1). Art. No.:CD003974.
[109] Roman GC, Salloway S, Black SE, Royall DR, DeCarli C, Weiner MW, Moline M, Kumar D, Schindler R, Posner H. Randomized, placebo-controlled, clinical trial of donepezil in vascular dementia: diffrential effects by hippocampal size. Stroke 2010;41:1213–21.
[110] Hyde AJ, May BH, Xue CC, Zhang AL. Variation in placebo effect sizes in clinical trials of oral interventions for management of the behavioural and psychological symptoms of dementia (BPSD): a systematic review and meta-analysis. Am J Geriatr Psychiatry 2017;25:994–1008.
[111] Pecina M, Heffernan J, Wilson J, Zubieta JK, Dombrovski AY. Prefrontal expectancy and reinforcement-driven antipressant placebo effects. Transl Psychiatry 2018;8:222. https://doi.org/10.1038/s41398-018-0263-y.
[112] Rutherford BR, Wall MM, Brown PJ, Choo T-H, Wager TD, Peterson BS, Chung S, Kirsch I, Roose SP. Patient expectancy as a mediator of placebo effects in antidepressant clinical trials. Am J Psychiatry 2017;174:135–42. https://doi.org/10.1176/appi.ajp.2016.16020225.
[113] Karnieli-Miller O, Werner P, Aharon-Peretz J, Sinoff G, Eidelman S. Expectations, experiences, and tensions in the memory clinic: the process of diagnosis disclosure of dementia within a triad. Int Psychogeriatr 2012;24:1756–70.
[114] Morgan DG, Walls-Ingram S, Cammer A, O'Connell ME, Crossley M, Bello-Haas VD, Forbes D, Innes A, Kirk A, Stewart N. Informal caregivers' hopes and expectations of a referral to a memory clinic. Soc Sci Med 2014;102:111–8.
[115] Jurgens FJ, Clissett P, Gladman JRF, Harwood RH. Why are family carers of people with dementia dissatisfied with general hospital care? A qualitative study. BMC Geriatr 2012;12:57.
[116] Stadnyk RL, Jurczak SC, Johnson V, Augustine H, Sampson RD. Effects of the physical and social environment on resident-family member activities in assisted living facilities for persons with dementia. Sr Hous Care J 2013;21:36–52.
[117] Read ST, Toye C, Wynaden D. Experiences and expectations of living with dementia: a qualitative study. Collegian 2017;24:427–32.

[118] McCleary L, Thompson GN, Venturato L, Wickson-Griffiths A, Hunter P, Sussman T, Kaasalainen S. Meaningful connections in dementia end of life care in long term care homes. BMC Psychiatry 2018;18:307.
[119] Adler G. Dementia caregivers: expectations of respite care. Am J Alzheimer's care Relat Disord Res 1992;7(2):8–12.
[120] Lee EA. Growing democracy through citizen involvement: a framework for measuring success. MSc thesis. Vancouver: School of Community and Regional Planning. University of British Columbia; 2017.
[121] Koehn S, Donahue M, Feldman F, Drummond N. Fostering trust and sharing responsibility between statutory and community sectors to increase access to dementia care for immigrant older adults. Ethn Health 2019. https://doi.org/10.1080/13557858.2019.1655529.
[122] Dixon-Woods M, Cavers D, Agarwal MS, Annandale E, Arthur T, Harvey J, Hsu R, Katbamna S, Olsen R, Smith L, Riley R, Sutton AJ. Conducting a critical interpretive synthesis of the literature on access to healthcare by vulnerable groups. BMC Med Res Methodol 2006;6(35). 07/27/06, https://doi.org/10.1186/1471-2288-6-35.
[123] Ruisoto P, Contador I, Fernandez-Calvo B, Palenzuela D, Romos F. Exploring the association between optimism anmd quality of life among informal caregivers of persons with dementia. Int Psychogeriatr 2018. https://doi.org/10.1017/S104161021800090W.
[124] Contador I, Fernandez-Calvo B, Palenzuela DL, Migueis S, Ramos F. Prediction of burden in family caregivers of patients with dementia: a perspective of optimism based on generalized expectancies of control. Aging Ment Health 2012;16:675–82.
[125] The WHOQOL Group. Development of the world health organozation WHOQOL-BREF quality of life assessment. Psychol Med 1998;28:551–8.
[126] Contador I, Fernández-Calvo B, Palenzuela DL, Campos FR, Rivera-Navarro J, de Lucena VM. A control-based multidimensional approach to the role of optimism in the use of dementia day care services. Am J Alzheimer's Dis Other Dementias November 2015;30(7):686–93.
[127] Ploeg J, Northwood M, Duggleby W, McAiney CA, Chambers T, Peacock S, Fisher K, Ghosh S, Markle-Reid M, Swindle J, Williams A. Caregivers of older adults with dementia and multiple chronic conditions: exploring their experiences with significant changes. Dementia March 6, 2019. https://doi.org/10.1177/1471301219834423.
[128] Bandura A. Self-efficacy: towards a unifying theory of behavioural change. Psychol Rev 1977;84:191–215.
[129] Fortinsky RH, Kercher K, Burant CJ. Measurement and correlates of family caregiver self-efficacy for managing dementia. Aging Ment Health 2002;6:153–60.
[130] Nogales-Gonzales C, Romero-Moreno R, Losada A, Marquez-Gonzalez M, Zarit SH. Moderating effect of self-efficacy on the relation between behavior problems in persons with dementia and the distress they cause in caregivers. Aging Ment Health 2015;19:1022–30. 10/1080/13607863.2014.995593.
[131] Rotter JB. Generalized expectancies for internal versus external control of reinforcement. Psychol Monogr Gen Appl 1966;80:1–28. https://doi.org/10.1037/h0092976.
[132] Petriwsky A, Parker D, O'Dwyer S, Moyle W, Nucidora N. Interventions to build resilience in family caregivers of people living with dementia: a comprehensive systematic review. JBI Database System Rev Implement Rep 2016. https://doi.org/10.11124/JBISRIR-2016-002555.

[133] Nordtug B, Krokstad S, Holen A. Personality features, caring burden and mental health of cohabitants of partnrs with chronic obstructive pulmonary disease or dementia. Aging Ment Health 2011;15:318–26. https://doi.org/10.1080/13607863.2010.519319.

[134] Bruvik FK, Ulstein ID, Ranhoff AH, Engedal K. The effect of coping on the burden in family carers of persons with dementia. Aging Ment Health 2013;17:973–8.

[135] Polenick CA, Struble LM, Stanislawski B, Turnwald M, Broderick B, Gitlin LN, Kales HC. "The filter is kind of broken": family caregivers' attributions about behavioral and psychological symptoms of dementia. Am J Geriatr Psychiatry 2018;26:548–56. https://doi.org/10.1016/j.jagp.2017.12.004.

[136] O'Rourke HM, Duggleby W, Fraser KD, Jerke L. Factors that affect quality of life from the perspective of people with dementia: a metasynthesis. J Am Geriatr Soc 2015;63(1):24–38.

CHAPTER 2

"It is not a disease, only memory loss": exploring the complexity of access to a diagnosis of dementia in a cross-cultural sample

Sharon Koehn[1,2]
[1]Gerontology, Simon Fraser University, Vancouver, BC, Canada; [2]Sharon Koehn Research Consulting, Vancouver, BC, Canada

Introduction

Accessing medical care and social supports for the person with dementia and their care partners is a complex undertaking that requires the alignment of a diverse array of resources at the individual, social, and organizational levels. The Candidacy Framework developed by Dixon-Woods et al. [1] provides a valuable lens through which the constituent processes of this undertaking become more apparent. This in turn facilitates identification of potential solutions to inhibitors of access at each level. To illustrate these processes, this chapter will draw on examples from a comparative study of the prediagnosis period of Alzheimer's Disease and Related Dementias (ADRD) in Canada's four major ethnolinguistic groups across four cities: Vancouver, Calgary, Ottawa, and Toronto [2–6]. The goal of this chapter is to provide a comprehensive understanding of the challenges to securing a diagnosis of dementia in Canada and to consider relevant health and social policy implications.

"Pathways to Diagnosis" study

The Social Science and Humanities Research Council (SSHRC) funded Pathways to diagnosis study was one of many inquiries into care for persons with dementia and their care partners, conducted by the pan-Canadian, interdisciplinary DementiaNET research group. The study focused on the initial identification of a problem, which at some point came to be labeled "dementia," and the decisions and experiences associated with seeking help

Evidence-informed Approaches for Managing Dementia Transitions
ISBN 978-0-12-817566-8
https://doi.org/10.1016/B978-0-12-817566-8.00002-4

about it. The team identified the following research questions: how and when do persons with dementia and their care partners identify a problem that becomes diagnosed as dementia; who do they consult for help; what are people's experiences in the prediagnosis period; what are people's expectations of the illness and its care; and what socially constructed meanings and values are embodied in respondents' discourse?

Data collection

To explore this transition in the context of the diverse ethnic, cultural, and linguistic backgrounds of older Canadians, the Pathways study targeted four sites, each focusing on one of the main ethnolinguistic groups in the country [7]: Anglo-Canadians in Calgary, Francophone-Canadians in Ottawa, South Asian-Canadians in Greater Toronto and Chinese-Canadians in Greater Vancouver. Inclusion criteria for persons with dementia were that they were aged 55 years or older; self-identified as belonging to one of our targeted groups; had received a diagnosis of Alzheimer disease, vascular, or mixed dementia in the preceding 4 years; were able to recall their prediagnostic experiences; provided informed consent; and had a family care partner willing to participate. In-depth interviews were conducted with 29 persons with dementia and 34 care partners. We interviewed persons with dementia—care partner dyads (sometimes triads, if they had more than one caregiver) based on evidence that collection of experiential data from persons with dementia is possible and contributes to a full understanding of the illness progression [8], while care partners are typically able to compensate for the dementia-impeded recollection of persons with dementia themselves. Participant characteristics are summarized in Figs. 2.1 and 2.2 and discussed in greater depth elsewhere [3—6].

In each community, participants were recruited by a research associate who was familiar with the target population and fluent in the relevant languages. Referrals to the study came from community agencies, memory clinics, and local branches of the Alzheimer Society of Canada. Approval was received from research ethics boards of all recruitment sites and investigators' universities.

Theoretical frameworks: candidacy for care and intersectionality

Critical, interpretive synthesis of access to care for "vulnerable" populations by Dixon-Woods et al. [1] included seniors and immigrants, among others, and was guided by a focus on equity. Its goal was to understand how access

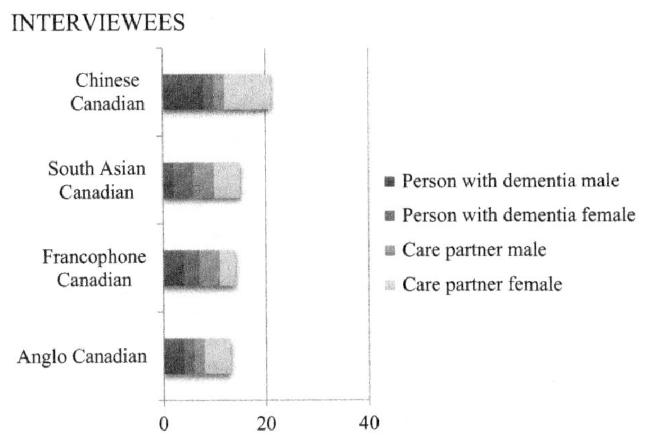

Figure 2.1 Gender distribution of persons with dementia and care partners.

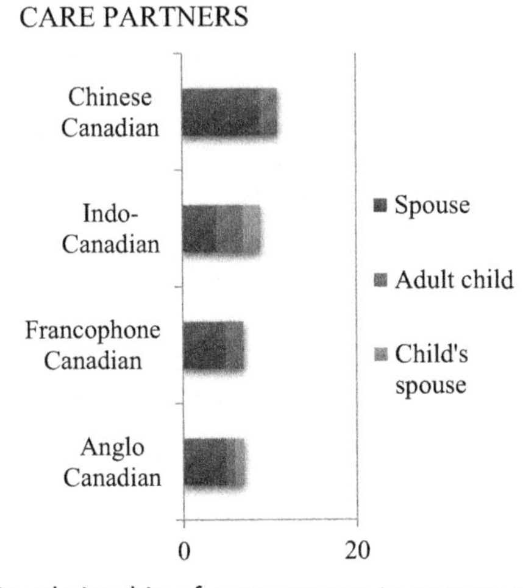

Figure 2.2 Kin relationship of care partners to persons with dementia.

is achieved by these groups within a health system, such as those in the United Kingdom and Canada, in which most care is free at the point of access. The complex construct of access is partitioned into seven dimensions: The first six of these are transition points at which a person's candidacy for care must be negotiated; the seventh captures the broader environmental context of negotiations (Table 2.1).

The Candidacy model of access to care thus speaks to a person's dynamic and continually negotiated sense of legitimacy in using healthcare [9]. Its dimensions illuminate inequities in health and healthcare "by tying seemingly individual behaviors in utilization to socially patterned influences" [9]. Increasingly, it is recognized as valuable for understanding

Table 2.1 The seven dimensions of access identified in the Candidacy framework.

Candidacy dimensions	Characteristics
Identification of candidacy	• Differential recognition of symptoms as needing medical attention. • Vulnerable populations are more likely to manage health as a series of crises. • Evidence of lower use of preventive services and higher use of accident and emergency facilities, emergency admissions and out-of-hours use" [1].
Navigation	• Awareness of the services on offer; known to be reduced for vulnerable populations. • Mobilization of practical resources—e.g. time off work, transportation—that are typically less readily available to vulnerable populations
Permeability of services	• Services are more or less accessible ("permeable") depending on the qualifications of candidacy required to use them (e.g. a referral) and the degree to which resources need to be organized. • Less permeable services "demand a higher degree of cultural alignment between themselves and their users" [1].
Appearances at health services	• Credibility once the client has presented at a health service depends on his/her competence in formulating and articulating the issue for which help is being sought
Adjudications	• Judgment calls made by the health professionals that clients initially consult. • "Professional perceptions of the cultural and health capital required to *convert* a unit of health provision into a given unit of health gain may function as barriers to healthcare. ... In addition, perceptions of social "deservingness" may play a role" [1].
Offers and resistance	• Resistance by patients to referrals and/or offers of medication.
Operating conditions	• "Locally specific influences on interactions between practitioners and patients" • "The perceived or actual availability and suitability of resources to address [a claim to] candidacy" [1].

the complex process of obtaining care in disadvantaged populations across diverse healthcare domains [9–15].

The diversity found within older adult populations that is reflected in our sample further demands that we account for difference without essentializing any given marker of identity. Intersectionality theorists maintain that while disparities arising from biological sex differences, gender, ethnicity or class affect well-being, relationships are dynamic; intersections and compounding of these dimensions affect health and quality of life as people age [16–18]. Being mindful of intersections is essential to avoid "culturalist explanations of illness and caregiving behaviors that (a) risk framing ethnocultural minorities as inadequate 'others' and (b) fail to account for power imbalances arising from intersections between multiple dimensions of difference" [3].

Initial problem identification and the peridiagnostic period

A retrospective analysis of the Pathways data from all four study sites through the Candidacy Framework lens provides important insights into the negotiations that people engage in with themselves and their families, as well as the healthcare system, from the point of recognizing that something is wrong to the attainment of a diagnosis. It also sheds light on societal beliefs and organizational configurations that mitigate or imperil these negotiations.

Our deductive thematic analysis used a qualitative descriptive approach [19]. Interview transcripts from all four communities were imported into NVivo 10 for coding. All relevant data in the 63 interviews were coded relative to the seven dimensions of the framework and inductively as subcategories of each code. Multiple levels of subcodes emerged within categories. Analytic emphasis was placed on experiences leading up to and including the diagnosis of dementia, as well as immediately afterward. This allowed us to consider how participants accepted or rejected offers of treatment, including medication and referral.

Dimensions of candidacy

Identification

The first step in gaining access to health care is determining that one needs and deserves it [1]. As in Krull's study [20], pivotal events that could no longer be interpreted as normal typically led to help-seeking in our sample. Examples include memory impairments resulting in difficulties with instrumental activities of daily living (e.g., medication management, managing money), impaired judgment (e.g., dressing for the wrong season), physical symptoms (e.g., shuffle walking, endless appetite), or radical behavior changes such as extreme apathy or aggression. More extreme examples are behaviors that gave rise to safety concerns, such as getting lost, falls, unsafe driving, causing flooding or fires, and inability to recognize rotten food. For most dyads, however, symptoms exhibited by the person with dementia were only gradually recognized as warranting medical attention through a process of appraising their abilities and skills relative to expectations of aging [21,22].

Persons with dementia in the study exhibited significant ambivalence about the implications of their memory loss and potential dementia diagnosis; in some cases, symptoms were simultaneously acknowledged and resisted [23]. Some persons with dementia recognized problems with cognition. For example, a Chinese-Canadian person with dementia went to consult a primary care physician because he was "*feeble-minded,*" a South Asian-Canadian person with dementia described cooking mishaps (burning food) and forgetting words, an Anglo-Canadian person with dementia was troubled by her lost ability to paint, and a Francophone-Canadian person with dementia noticed his slower reflexes and tendency to make more driving errors. Simultaneously, however, persons with dementia were also likely to deny or normalize symptoms: A Chinese-Canadian person with dementia said, *"[I]t is not a disease, only memory loss,"* and a South Asian person with dementia said, *"[T]his is the way of the world .... (laughs) There's nothing special about this."* Others tried to hide or deny their disease: The wife of a South Asian-Canadian person with dementia said, "*Not even the children know about it. He's very secretive in that respect.*" An Anglo-Canadian person with dementia reflected, *"I probably was in denial in some ways, saying I'm okay."* Anglo-Canadian persons with dementia were unique in noticing their problems before their care partners and playing a key role in independently seeking help: *"[A]s soon as he knew what his diagnosis was, like he's been on the internet. This guy's a fighter. He's determined. He wants to beat this"* (Anglo-Canadian care partner). It was noted that in the Anglo-Canadian sample "because participants were sampled from a peer-support program, potentially, they may be more proactive ... and may not experience the disease passively" [4].

Family care partners commonly attributed symptoms such as forgetfulness to "normal aging" or "human nature." Care partners in each of the Chinese- and South Asian-Canadian samples attributed symptoms to the person with dementia's naturally reserved or "odd" personality. With the exception of those in the Chinese-Canadian sample, care partners confused the symptoms of dementia with other illnesses (particularly depression) or attributed them to comorbid disorders or to medications taken for them. A Francophone-Canadian care partner attributed her mother's failing memory to her "inadequate" self-care (nutrition and exercise) rather than to the root cause of both, dementia. An Indo-Canadian care partner offered a culturally informed interpretation of the breakdown of his wife's cognitive ability, attributing it to the Law of Shiva whereby nature is continuously transformed and dissolved. Regardless of their expression, misinterpretations of symptoms by care partners could be traced back to a lack of knowledge about the signs and symptoms of dementia, a point that care partners were quick to make. For example, *"I have a certificate in gerontology and I know nothing about that [dementia] .... It's been a long time, in the 80s .... I took all the courses. But you know, it's unbelievable"* (Francophone-Canadian care partner).

Family care partners who were quicker to recognize symptoms and more proactive in seeking medical attention had greater awareness of dementia. Having prior experience of the condition helped to facilitate this recognition; seeing the person with dementia after an extended period of time also triggered recognition by some family members. For example, a retired Chinese-Canadian nurse told us, *"[I]f I am not from medical field, I would [be] like anybody else, not aware of his changes."* Care partners were exposed to information about dementia on television, in magazines, and at ethnolinguistically targeted presentations at churches. Among help-seeking Chinese-Canadian samples, some care partners actively sought information about memory loss and dementia online. Such activities facilitated the recognition of dementia as distinct from normal aging or something else.

Navigation

Once they have established a need for services, the person with dementia and/or the family care partner need to find them. This requires awareness of the services on offer as well as a means of mobilizing resources, such as language, transportation, and social supports. Proactive care partners with knowledge of the healthcare system and the education and confidence to seek out information and advocate for the needs of the person with dementia were invaluable as navigators. Care partners fluent in English or French told us that they accessed online resources to find supports they needed. Alzheimer Society resources, including a Chinese support group, were most commonly cited. For those less able to seek out mainstream resources, linking organizations were crucial. The Alzheimer Society supplied referrals to other resources and information, but some needed additional linkages to access the Alzheimer Society itself. A community agency serving the Chinese population in Vancouver partners with the Alzheimer Society of British Columbia to provide space for monthly support group meetings, caregivers' workshops, and other educational workshops for their Chinese-speaking clients and the general public. In addition, they support an annual Chinese forum on brain health and dementia. In return, the Alzheimer Society of British Columbia provides training workshops to their staff and volunteers who work closely with clients affected by the disease. In Toronto, a community-care case manager referred a dyad to a South Asian community health and social service agency that was able to provide ethnolinguistically appropriate support.

Family members outside of the person with dementia—care partner dyad assisted navigation at differing levels of involvement. Some merely encouraged the dyad to seek care or gave advice, others arranged consultations with primary care physicians or specialists, and some accompanied either the person with dementia or the dyad to appointments. Most were sons or daughters of the dyad, but in some cases, the persons with dementia's or care partner's siblings and their family members (e.g., a niece who was a nursing student) took the initiative. Friends and neighbors were also important. Typically, persons with dementia rely on their family care partners, most often their spouses, for transportation to doctor's appointments and elsewhere, but adult children and nonfamily members also assume this role.

Navigation was thwarted by transportation or mobility issues, lack of awareness of service availability and how service could be utilized, family responsibilities or a lack of family support, and social environmental factors. Several participants commented that the offices of people they needed to consult were too far from their homes. A Francophone couple moved to Ottawa from an isolated region of Quebec when traveling to an urban center for health care proved too onerous. Taking transit was difficult for older South Asians who had no prior Canadian transit experience, got lost when attempting to do so, had difficulty getting to the bus stop in the winter months, and whose bus service did not run on Sundays. One person with dementia was "*too proud*" to use the specialized accessible public transportation service [24].

At least one care partner at each site reported not knowing about the Alzheimer Society or what resources it could provide. A Chinese-Canadian caregiving daughter felt disadvantaged by her lack of knowledge about available resources, having only been in Canada for 3 years. She said that *"the help is not in depth or adequate"* and *"sometimes, you don't know how to approach it at first, ...where to find the information."* The daughter of a South Asian-Canadian person with dementia commented, we are *"left on our own to find things. There may be services we don't access because we don't know about them."* Not knowing how to look things up on the internet was one problem; for non-English speakers, not having online access to information in their own language was another. Care partners cited a lack of time after work as a reason for not using the computer to locate services. Anglo- and Francophone-Canadian persons with dementia said they had "*given up*" using computers because they had "*lost interest*" or found it difficult.

Importantly, we also heard that family members are not always available or willing to help with the navigation process. Across all sites, we heard from couples who had no children or whose children were "*too busy*" or lived too far away or were not emotionally close to them. As a result, the person with dementia and caregiving spouse could not rely on adult offspring to accompany them to the doctor, give them a ride, or even provide moral support. A Francophone-Canadian caregiving daughter described the difficulty of a demanding job, being a single parent, and taking primary responsibility for her parent with dementia. A Chinese-Canadian caregiving daughter was a new immigrant with limited English and a new baby. In combination, these factors prevented her from finding the information to direct her mother to appropriate services. Even spousal navigational support cannot be taken for granted. A Chinese-Canadian spousal care partner was limited in the navigation assistance she could provide because she was also caring for a grandchild.

Some of our participants said that asking for help from others outside their inner circle of family and friends could be difficult. A couple who had recently moved to Ottawa only had new friends from whom they "*would never ask for anything.*" An Anglo-Canadian person with dementia was afraid of talking to people about dementia because he had noticed people's "*attitudes toward the disease*" and "*how they react to you.*" Another in the Chinese-Canadian sample told us that he avoids talking about his dementia to his fellow churchgoers for "*fear of giving it to them,*" or because it will "*scare people away,*" and a Francophone-Canadian person with dementia said she keeps her diagnosis to herself because "*maybe some will say: she's crazy!*" The experience of a Francophone-Canadian care partner indicates that these fears have some grounding in reality: "*We don't have too many friends left ... [people]*

withdraw from you! It's not a contagious disease!" These examples speak to limitations of social networks if they are not built on a foundation of trust and knowledge and to the pervasiveness of the stigma associated with dementia across different cultural communities. Both can impede navigation.

Help-seeking is also inhibited when people believe that no suitable options are available. For example, a care partner in the Anglo-Canadian sample believed there was scant support for dementia other than the Alzheimer Society and was tired of being referred back to the person with dementia's primary care physician who she felt was insufficiently knowledgeable about Alzheimer's Disease and related dementias. And a Chinese-Canadian care partner said, "*when you really need a person to help you out, apart from the family members ... there is nobody who can help.*"

Appearances and adjudications

The Candidacy framework [1] distinguishes two dimensions of the interaction between a patient and health professional that influence the ability to establish a need for care. The first, "Appearances," refers to the patient's competency in articulating the health issue or their need for care; language skills, social class, gender, age, and cognition are important influences. The second dimension refers to the judgments (or "Adjudications") of a patient's needs that are made by the health professionals based on their own knowledge, experiences, and biases. We analyzed Appearances and Adjudications simultaneously as they often appeared concurrently in participants' narratives. The visit to the primary care physician and the circumstances surrounding the establishment of a diagnosis are the two overarching themes that speak to these dimensions.

The outcome of visits to a primary care physician was influenced by the communication skills of patients and providers and by the relative influence of the person with dementia and their care partner in the encounter. This was influenced, in turn, by their beliefs and expectations. Overall, participants spoke more to the attributes of the physician's communication skills than to their own. One primary care physician of a participant in the Francophone-Canadian sample, described as good at explaining things, was deemed extremely helpful and contrasted specifically to a neurologist who was depicted as "*busy and very clinical.*" Conversely, Chinese-Canadian care partners referred to their Chinese-speaking doctors as relatively uncommunicative: One said nothing about the diagnosis, another withheld the results of a blood test, and a third did not mention the medication dosage. In the latter case, the care partner gave the person with dementia the wrong dose, which made him sick. Several Chinese- and South Asian-Canadian dyads reported that medication was prescribed without the care partner understanding what it was for (see also [25]). Similarly, a South Asian-Canadian person with dementia complained that the doctors "*never talk to me. They mention everything to my daughter.*" Similar to the patient, the primary care physician was fluent in Hindi. In these cases, the problem is physician communication style, not a language barrier.

The inability of the person with dementia to communicate his or her needs was identified by Francophone-Canadian participants. One care partner told us that the person with dementia *"doesn't chat as much,"* which required intensive questioning by the physician. Another Francophone-Canadian care partner thought that the person with dementia was afraid to communicate what she was experiencing for fear that she would be placed in an institution. These examples speak to the nature of the disease itself as an impediment to coherent presentation.

At least one person with dementia in each community had visited their primary care physician independently, usually to replenish medications for a comorbid condition, although two—among Anglo- and Francophone-Canadian participants—specifically presented their dementia symptoms to their primary care physician. Among South Asian-Canadian participants, we noted five instances whereby a person with dementia avoided or tried to avoid detection of dementia symptoms by visiting the primary care physician without the care partner, who they thought would disclose the person with dementia's difficulties. Their reluctance to disclose to their primary care physicians was related to beliefs that they did not need help, they did not know how to talk to their doctor, the doctor was not sufficiently qualified to help them, or the doctor would not disclose their condition to them to protect them from worry. A minority of Chinese- and South Asian-Canadian family care partners also believed that their doctor could not help them with the symptoms they witnessed (e.g., forgetfulness) because they did not associate them with a "disease": "*we will not tell them anything which is not related with the diseases, they will not help us!*" (Chinese-Canadian care partner).

Overall, however, family care partners were pivotal in drawing the primary care physician's attention to the person with dementia's symptoms, sometimes inciting the anger of a person with dementia who "*wanted to prove to the doctor that he was OK*" (South Asian-Canadian care partner). Many family care partners, particularly spouses, said they always accompanied the person with dementia to consultations and were present for the diagnosis. Most of them took this opportunity to voice their concerns over the changes they had witnessed in the person with dementia. For some, these reports led to the person with dementia's assessment or referral to a specialist, but in each community, care partners complained that their primary care physician was not receptive to their concerns and was slow to refer the person with dementia for dementia testing. Anglo- and Francophone-Canadian care partners said they had to get more "*insistent*" or "*demanding,*" but it appears that Chinese- and South Asian-Canadian interviewees felt more disempowered by the primary care physician's disregard of their concerns: "*the family doctor said 'no' in the beginning, he did not think of it. He felt I might be too sensitive*" (Chinese-Canadian care partner).

The presence of comorbidities often confounded diagnosis, although at times physicians would identify symptoms of ADRD when treating the patient for another ailment. For example, the sleeping problems of one person with dementia were attributed to restless leg syndrome, and other symptoms were thought to be side effects of the medication for that disorder. However, the need for emergency treatment after a traffic accident brought the dementia of another person with dementia to the attention of healthcare providers for the first time. Primary care physicians were also inclined to attribute dementia symptoms to "normal aging" or would diagnose and sometimes treat symptoms as disorders such as sleep apnea, depression, or adjustment disorder. Participants at all sites typically made at least three visits to a primary care physician or specialist before securing a diagnosis and recalled having multiple scans and tests.

Participants who recalled seeing specialists typically commented on their limited communication: *"He was in a hurry and he did not say anything and we dared not ask him. We took the prescription and went home*" (Chinese-Canadian person with dementia); however, some saw a lack of empathy as the problem: *"[the specialist] was very bad. He was not empathetic at all. He didn't even listen to our side"* (South Asian-Canadian care partner). Lack of communication also resulted in providers incorrectly assuming that an Anglo-Canadian person with dementia's husband was her primary care partner.

Offers and resistance

This dimension of Candidacy refers to the acceptance and resistance to offers of care [1], which may be influenced by the degree of alignment between the offer and the recipient's health beliefs, cultural suitability, as well as various social determinants of health [12]. In our analysis, participants' acceptance and resistance to offers were charted according to who made the offer (e.g., primary care physician or specialist) and what the offer was for.

Following interactions with the primary care physician, offers commonly included prescriptions, drug samples, follow-up appointments, referrals to memory specialists, and emotional support for care partners. Offers made by specialists included prescriptions and changes in medication, referrals to a new doctor, diagnostic scans, and drug samples. Offers were most often accepted on the basis of trust in the doctor. As recounted by a Chinese-Canadian person with dementia, *"I will take (medication) because he is a doctor. I believe my family doctor."* Correspondingly, resistance to offers was frequently attributed to lack of confidence in the doctor. For example, a South Asian-Canadian care partner questioned the suitability of medication because it was prescribed in the absence of diagnostic imaging.

Offers of community-based supports included education and peer programs at the local Alzheimer Society, adult day programs, personal support workers and social workers, housekeeping services, and ethnospecific community groups. Care partners and persons with dementia who accepted such offers appreciated the "continuity" of support and access to educational resources.

Yet, several participants in our study expressed disappointment at a "lack of offers." An Anglo-Canadian care partner noted, "*It's not just the testing, it was the support. That's what I find is that there is no support.*" Generally, care partners' resistance to offers was due to their perception that offered services or advice was not "*concrete*" or "*practical.*" While the care partners in our study mostly appreciated offers of educational sessions and care partner support groups, several thought they were unsuitable: "*I think that it's bad enough to live with a situation without having to listen to everyone else's problems*" (Francophone-Canadian care partner).

Resistance to offers by persons with dementia was generally due to worries about side effects from medications, the complexities of managing medications, or the desire to maintain their independence. The perceived stigma associated with diagnosis is reflected in the rejection of the offer of support groups by two persons with dementia who did not want to be "*put in with all the loonies*" (Francophone-Canadian) or exposed to *"people who were far worse*" (Anglo-Canadian) in terms of disease progression.

Interestingly, the decision to accept or refuse an offer was frequently influenced by family and cultural factors. In particular, shifting gender roles within dyads made it difficult for participants to accept support.

A Francophone-Canadian caregiving husband was reluctant to step away from his "*role as protector*" to obtain respite, whereas an Anglo-Canadian PWD struggled to accept her husband's new caregiving role. Among Chinese-Canadian participants, other family members were frequently brought into discussions about offers. Several South Asian-Canadian participants mentioned accepting offers in combination with complementary and alternative medicine.

Permeability of services

The organization of health services can render them more or less accessible or "permeable"; services have low permeability when they require a high degree of alignment between users and providers, the mobilization of many resources, and impose qualifications for eligibility [1]. Conditions that limit permeability include health literacy, language, hours of operation, and funding models for providers [12].

Three common themes relating to permeability emerged: (1) alignment with service users (e.g., preferences for certain provider characteristics), (2) aspects of the physical environment (e.g., office location), and (3) system-level factors (e.g., the referral process). These factors are especially significant for persons with dementia who often use more services during the prediagnosis period, when differences between dementia symptoms and normal aging are ambiguous [25].

The importance of alignment between participants and health services and its effect on service permeability was evident across all four sites. For instance, a Francophone-Canadian care partner told us that you have to try "*two or three different (psychologists) before you can find someone that you can click with.*" The account of an Anglo-Canadian person with dementia who surprised her caregiving son when she selected a male primary care physician after searching pointedly for a female one illustrates the complexity of identity and thus the task of finding good alignment. For long-time Canadians, such as those from the Francophone-Canadian community, alignment resulted in and was augmented by continuity with primary care physicians who were familiar with the person and able to monitor changes in their long-term health.

Not surprisingly, ethnolinguistic congruence between participants and services was important for Chinese-, South Asian-, and Francophone-Canadian dyads. Chinese-Canadian care partners and persons with dementia generally preferred services offered in Chinese languages. Certain Chinese-Canadian care partners noted the need for additional bilingual resources: One told us that some of her acquaintances experiencing Alzheimer Disease *"give up all the treatment because they cannot find the resources and strong support."* Similarly, a South Asian-Canadian person with dementia was worried about finding a care home in which there was "*someone with whom [she could] carry on a conversation.*" Ethnospecific linking organizations were pivotal to ensuring service permeability for Chinese- and South Asian-Canadian participants.

Ethnolinguistic congruence can also influence diagnosis, for which the physicians' advanced communication skills are critical. For example, a Francophone-Canadian person with dementia was reverting to a dialect from France and had difficulty communicating with his neurologist. The Canadian Government has funded two organizations that aim to improve access to health services by French speakers and increase the number of health professionals who offer services in French [6]; several Francophone care partners nonetheless remarked on the low permeability of physician services in their language.

Permeability was further affected by environmental factors. Several dyads with limited mobility (in walking and driving) reported difficulty getting to services located in distant parts of the city or on the upper floor of a building without an elevator. For example, a care partner said of a memory clinic, "*We'd get there; we'd park the car. It was complicated - at the end of the hallway. And I can't walk easily. At the far, far end of the hallway and all for nothing.*"

The organization of healthcare systems further influenced permeability. Our participants expressed frustration with long wait times to access services in the first place (2 weeks to 3 months) and to obtain subsequent test results or specialist appointments (6–8 months in one case). The process of securing appointments was another barrier. Participants remarked on referrals back and forth between linking organizations, specialists, and primary care physicians, increasing the time required. Several francophone dyads were fortunate to have a caseworker who regularly phoned and visited them, thus increasing permeability.

Discussion

The Candidacy Framework is valuable insofar as it unpacks the multiple and diverse challenges faced by people with dementia, and others in their lives, in identifying that they do in fact have a problem in need of medical attention, and then actually securing a diagnosis. Our cross-cultural sample deepens our understanding of how these processes play out in diverse ways for individuals and their care partners in their communities and interactions with the healthcare system. Also apparent from our data is the importance of considering intersections of identity and their influence in different contexts on social capital, which in turn has a positive correlation with access to needed services. With samples in four provinces, we also had an opportunity to hear from people who have encountered different configurations of service delivery. Thus while we cannot generalize too broadly from our qualitative findings, there are, nonetheless, some valuable lessons to be learned from the integrated analysis of our four case studies.

At the level of the individual, we identified a "fluctuating awareness" [26] or normalization [27] among persons with dementia of their symptoms that may be seen as self-protective, a means by which they cope with the cognitive and social changes of dementia. As Sabat [28] argues, persons with dementia often emphasize their mentally competent selves when treated as someone who is no longer "normal" by healthy others who threaten their sense of agency and self-hood. This argument is consistent with Bourdieu's position that economic, cultural, and symbolic capital available to people in different social spaces affords them different levels of (symbolic) power [29]. Persons with dementia are typically disadvantaged in this regard because their cognitive impairment erodes their access to, and command of, social capital. This increasingly depletes with disease progression [28]. The role reversals that dementia can impose on spouses, accompanied for some by a loss of power in the relationship, also generated resistance to the care partners' assistance by some persons with dementia. Intersections of identity (i.e., being white, English-speaking, well-educated) also influence social capital [16,17]. As members of a dominant majority with strong English language skills, the Calgary-based Anglo-Canadian persons with dementia were better positioned to access information and were more inclined to approach their physicians independently.

Persons with dementia and care partners across different cultures were also hesitant to seek help from others because of the stigma associated with dementia. While the stigma and shame associated with cognitive impairment has been foregrounded in studies of people of "Asian" cultural background [30–33], our study revealed more instances of stigmatization in the "non-Asian" samples from the Anglo- and Francophone-Canadian communities [34]. This internalized stigma is also apparent in the rejection by some persons with dementia of support groups offered by the Alzheimer's Society.

Among persons with dementia and care partners alike, lack of knowledge hindered help-seeking across different ethnocultural groups. These findings are congruent with the literature. For example, 70% of the care partners in the European survey of Bond et al. [35] attributed a lag time in help-seeking to this reason. Immigrant populations with little exposure to educational media throughout their lives may be further disadvantaged in this respect [36], as are minorities with inadequate access to information on dementia in their mother tongue [37,38]. Several studies have identified limited knowledge of dementia as a barrier to identification among South Asians in the United Kingdom [30,39,40]. Among Francophones in Canada, improved health literacy is needed to enhance access to health services [41].

Trusted care partners proved essential in terms of problem identification, navigation to services, and as advocates for the person with dementia's needs in interactions with health service providers and other gatekeepers. As with persons with dementia, however, it is important to consider intersections of identity of the care partners. Higher levels of social capital within the person with dementia's family facilitated access to knowledge about dementia symptoms and hence the ability to correctly interpret the person with dementia's behaviors. Being an immigrant is not necessarily a disadvantage if the person has aged in Canada, speaks English, and either worked as a health professional or has family or friends who do. Care partners with higher levels of social capital were also better able to identify the services needed and navigate to them. English language capability is important, but so too is familiarity with the system (knowing what to look for and how to find it). Health and computer literacy are valuable skills in this regard [42].

Notably, however, several of the persons with dementia in our sample across different ethnic groups had no access to a care partner who could or would provide the support they needed to secure a diagnosis or assistance with accessing care. Some were childless; others spoke of children or even spouses who were geographically or emotionally unavailable. The competing demands of caring for children or grandchildren can also limit the ability of adult children or spouses to assist persons with dementia. In line with Keefe [43], our findings show that structural factors relating to the availability of family members and to the patterning of helping behavior by gender may in fact be more important than ethnicity or filial piety in determining whether elders receive navigational support from family members.

For newer immigrants, or those with limited majority-language skills, the importance of "linking organizations" that connect formal and community-based services emerged. Other research corroborates that community-based agencies, particularly those from the multicultural and settlement sectors, are often instrumental in reaching out to immigrant and at-risk communities with health information [44] and navigational support [45,46]. Immigrant agency staff are able to leverage identification-based trust, based on their shared values and identity with their clients and the emotional ties they develop with them [47]. Building on evidence for their effective utilization in cancer care [44—46], a strong case may also be made for the development of culturally and linguistically competent patient navigators responsive to the early signs, symptoms and initial care-seeking and ongoing system navigation of people with dementia [48-50].

In contrast, a reticence to consult professional services reflects the limited trust that people often have in these providers. The finding of Bebbington et al. [51] that high levels of underconsultation with mental health professionals reflects low levels of confidence in their efficacy is reflected in the accounts of some of our participants. Especially salient in their accounts was the communication style of the care provider. Primary care physician communication skills are important to the management of dementia [52], and physicians' limited communication with and support of family care partners has been criticized [50]. Interactions between primary care physicians and patients are challenged by time constraints [53,54]. Deception of patients concerning their dementia diagnosis has also been justified by medical students in terms of the patient's capacity to understand, concern for the patient's vulnerability and family dynamics, and the physician's doubts about their own ability to communicate effectively [55]. Such concerns may be amplified by a culturally influenced reluctance on the part of the physician to share a diagnosis of dementia directly with the patient [27,44].

On the one hand, we found that persons with dementia and their care partners sought out care providers with congruent identities, particularly with respect to language and especially when persons with dementia reverted to their mother tongue [56]. Language barriers, lack of culturally appropriate resources, and discrimination have all been identified as barriers to diagnosis and additional care [57,58]. Fitzpatrick and Vangelisti [59] maintain that trust and understanding will best develop in healthcare interactions between members of similar groups, where patients may feel most comfortable interacting with health professionals of a similar identity (i.e., gender, age, class, ethnicity). What our data reveal, however, are that identities are complex and congruity on one dimension does not guarantee congruity on another. For example, poor communication was not limited to situations of language incongruence between the person with dementia-care partner dyad and the care provider; in fact, several of our Chinese and South Asian participants complained that language-congruent providers either withheld information, provided inaccurate information or elected to ignore the person with dementia, and speak only to their adult children in English. At two sites, we saw evidence of gender-based power imbalances, possibly enhanced by the physicians' professional status, when female care partners' concerns about the as yet undiagnosed persons with dementia were dismissed [3]. Ensuring language congruence is thus a necessary but not a sufficient measure to ensure good communication.

Also contentious for our participants was the seeming inability or reluctance of the primary care provider to confirm a diagnosis, or to clearly indicate what treatment was needed. Numerous studies have identified underdiagnosis of dementia by primary care physicians as a problem, with as few as 50% of dementia cases diagnosed by them [60,61]. A range of complex factors contribute to this, including the subtlety and variability of dementia symptoms, confounding comorbidities, the limited time primary care physicians have to spend with the patient, negative attitudes toward the importance of assessment and diagnosis, and lack of a definitive diagnostic test [62,63]. Primary care physicians are least likely to accurately diagnose dementia when patients have low education, low literacy, and are non-English-speaking [60,62].

Participants also expressed concerns about the limited information provided to them by their primary care physicians concerning community-based and/or psychosocial supports after a diagnosis. Previous research suggests that although persons with dementia and their families rely on their primary care physician for access to care, physicians may not think they have the training or time to explore these resources with patients [52,53].

Conclusions

Identifying signs of dementia and establishing a diagnosis is achieved through multiple and at times simultaneous negotiations. These occur internally for the person with dementia and between persons with dementia, their care partners, and healthcare providers. The Candidacy Framework establishes how these negotiations require the alignment of resources between the parties to the negotiation. These include knowledge of dementia and services available, health and computer literacy, language competencies, and cultural understandings, to name a few. Implied but not foregrounded in the Candidacy Framework is the critical importance of imbalances in social capital or power between the parties to the negotiation. Persons with dementia may try to hide symptoms or reject help to avoid losing power in their close relationships, and female care partners struggle to convince more powerful male physicians of the validity of their claims that their spouse's behavior is abnormal.

Social capital is also important insofar as it facilitates access to knowledge and hence the power to identify symptoms, identify, and navigate to resources and assess and effectively utilize offers, such as screening tests, diagnosis, medication, and psychosocial supports. Variations in social capital existed across all of the ethnocultural groups included in our sample, and these differences rather than ethnicity or culture per se, governed the ease with which they obtained a diagnosis and access to related services.

Identity congruency is deemed to be important to the establishment of trust between the person with dementia-care partner dyad on the one hand and formal care providers on the other. Ethnolinguistic congruency receives the most attention, but our findings attest to the importance of considering intersections of different dimensions of identity such as gender and culture, particularly in relation to the power these confer in relationships. Identity congruence and a genuine concern for communication by care providers can foster trust, which is deemed to be essential to overcoming power imbalances that thwart symptom identification and diagnosis. Trust in primary care physicians further depends on their ability to provide timely and accurate information and appropriate referrals. Systemic barriers such as short appointment times with primary care physicians and long wait times for specialists and screening can simultaneously undermine trust in the system and in primary care physicians whose ability to attend to relational care is thus limited. A shift to more patient-centered or relational care is supported by this analysis.

References

[1] Dixon-Woods M, Cavers D, Agarwal MS, Annandale E, Arthur T, Harvey J, et al. Conducting a critical interpretive synthesis of the literature on access to healthcare by vulnerable groups. BMC Med Res Methodol 2006;6(35).

[2] Koehn S, Badger M, McCleary L, Cohen C, Drummond N. Negotiating access to a diagnosis of dementia: implications for policies in health and social care. Dement Int J Soc Res Pract 2016;15(6):1436–56.

[3] Koehn S, McCleary L, Garcia L, Spence M, Jarvis P, Drummond N. Understanding Chinese-Canadian pathways to a diagnosis of dementia through a critical-constructionist lens. J Aging Stud 2012;26(1):44–54.
[4] Leung K, Finlay J, Silvius J, Koehn S, McCleary L, Cohen C, et al. Pathways to diagnosis: exploring the experiences of problem recognition and obtaining a dementia diagnosis among Anglo-Canadians. Health Soc Care Community 2011;19(4):372–81.
[5] McCleary L, Persaud M, Hum S, Pimlott NJG, Cohen CA, Koehn S, et al. Pathways to dementia diagnosis among South Asian Canadians. Dement Int J Soc Res Pract 2013;12(6):769–89.
[6] Garcia L, McCleary L, Emerson V, Leopoldoff H, Dalziel W, Drummond N, et al. It's all the little things: the pathway to diagnosis for certain Francophones living in a minority situation. Gerontol 2014;54(6):964–75.
[7] Statistics Canada. 2011. Census of population: linguistic characteristics of Canadians. 2012. Available from: http://www.statcan.gc.ca/daily-quotidien/121024/dq121024a-eng.htm.
[8] Phillipson L, Hammond A. More than talking: a scoping review of innovative approaches to qualitative research involving people with dementia. Int J Qual Methods 2018. https://doi.org/10.1177/1609406918782784.
[9] Klassen AC, Smith KC, Shariff-Marco S, Juon H-S. A healthy mistrust: how worldview relates to attitudes about breast cancer screening in a cross-sectional survey of low-income women. Int J Equity Health 2008;7(1):5–24.
[10] Abbott P, Magin P, Davison J, Hu W. Medical homelessness and candidacy: women transiting between prison and community health care. Int J Equity Health 2017;16(1):130. https://doi.org/10.1186/s12939-017-0627-6.
[11] Hunter C, Chew-Graham C, Langer S, Stenhoff A, Drinkwater J, Guthrie E, et al. A qualitative study of patient choices in using emergency health care for long-term conditions: the importance of candidacy and recursivity. Patient Educ Couns 2013;93(2):335–41.
[12] Koehn S. Negotiating candidacy: ethnic minority seniors' access to care. Ageing Soc 2009;29(4):585–608.
[13] Macdonald S, Blane D, Browne S, Conway E, Macleod U, May C, et al. Illness identity as an important component of candidacy: contrasting experiences of help-seeking and access to care in cancer and heart disease. Soc Sci Med 2016;168:101–10.
[14] Kovandžić M, Chew-Graham C, Reeve J, Edwards S, Peters S, Edge D, et al. Access to primary mental health care for hard-to-reach groups: from 'silent suffering' to 'making it work. Soc Sci Med 2011;72(5):763–72.
[15] Bristow K, Edwards S, Funnel E, Fisher L, Gask L, Dowrick C, et al. Help seeking and access to primary care for people from "hard-to-reach" groups with common mental health problems. Int J Fam Med 2011. https://doi.org/10.1155/2011/490634. Online(Article ID 490634):10pp.
[16] Hankivsky O. Health inequities in Canada: intersectional frameworks and practices. Vancouver, BC: UBC Press; 2011.
[17] Hulko W. Intersectionality in the context of later life experiences of dementia. In: Hankivsky O, editor. Health inequities in Canada: intersectional frameworks and practices. Vancouver: UBC Press; 2011. p. 198–217.
[18] Warner DF, Brown TH. Understanding how race/ethnicity and gender define age-trajectories of disability: an intersectionality approach. Soc Sci Med 2011;72(8):1236–48.
[19] Boyatzis RE. Transforming qualitative information: thematic analysis and code development. Thousand Oaks, CA: Sage; 1998.
[20] Krull AC. First signs and normalizations: caregiver routes to the diagnosis of Alzheimer's disease. J Aging Stud 2005;19(4):407–17.

[21] Campbell S, Manthorpe J, Samsi K, Abley C, Robinson L, Watts S, et al. Living with uncertainty: mapping the transition from pre-diagnosis to a diagnosis of dementia. J Aging Stud 2016;37:40–7.
[22] Groen-van de Ven L, Smits C, Oldewarris K, Span M, Jukema J, Eefsting J, et al. Decision trajectories in dementia care networks: decisions and related key events. Res Aging 2017;39(9):1039–71.
[23] Macquarrie CR. Experiences in early stage Alzheimer's disease: understanding the paradox of acceptance and denial. Aging Ment Health 2005;9(5):430–41.
[24] Toronto Transit Commission. Using wheel trans service. 2018. Available from: http://www.ttc.ca/WheelTrans/About_Wheel-Trans/index.jsp.
[25] O'Donoughue Jenkins L. Cognitive impairment and service use: the relationship between research and policy (Dissertation). Canberra, Australia: Australian National University; 2017. Available from: https://openresearch-repository.anu.edu.au/bitstream/1885/146632/1/Lily O'Donoughue Jenkins thesis 2018.pdf.
[26] Phinney A. Fluctuating awareness and the breakdown of the illness narrative in dementia. Dementia 2002;1(3):329–44.
[27] Clare L. We'll fight it as long as we can: coping with the onset of Alzheimer's disease. Aging Ment Health 2002;6(2):139–48.
[28] Sabat SR. Surviving manifestations of selfhood in Alzheimer's disease: a case study. Dementia 2002;1(1):25–36.
[29] Siisiainen M. Two concepts of social capital: Bourdieu vs. Putnam. Int J Contemp Sociol 2003;40(2):183–204.
[30] Giebel CM, Zubair M, Jolley D, Bhui KS, Purandare N, Worden A, et al. South Asian older adults with memory impairment: improving assessment and access to dementia care. Int J Geriatr Psychiatry 2015;30(4):345–56.
[31] Bowes A, Wilkinson H. "We didn't know it would get that bad": South Asian experiences of dementia and the service response. Health Soc Care Community 2003;11(5):387–96.
[32] Mackenzie J. Stigma and dementia: East European and South Asian family carers negotiating stigma in the UK. Dementia 2006;5(2):233–47.
[33] Zhan L. Caring for family members with Alzheimer's disease: perspectives from Chinese American caregivers. J Gerontol Nurs 2004;30(8):19–29.
[34] Evans SC. Ageism and dementia. In: Ayalon L, Tesch-Römer C, editors. Contemporary perspectives on ageism. Cham: Springer; 2018. p. 263–75. Available from: https://link.springer.com/book/10.1007/978-3-319-73820-8.
[35] Bond J, Stave C, Sganga A, Vincenzino O, O'connell B, Stanley R. Inequalities in dementia care across Europe: key findings of the facing dementia survey. Int J Clin Pract 2005;59(s146):8–14.
[36] Ferri CP, Prince M, Brayne C, Brodaty H, Fratiglioni L, Ganguli M, et al. Global prevalence of dementia: a delphi consensus study. Lancet 2005;366(9503):2112–7.
[37] Fornazzari L, Fischer C, Hansen T, Ringer L. Knowledge of Alzheimer's disease and subjective memory impairment in Latin American seniors in the greater Toronto area. Int Psychogeriatr 2009;21(5):966–9.
[38] Low L-F, Anstey KJ, Lackersteen SM, Camit M, Harrison F, Draper B, et al. Recognition, attitudes and causal beliefs regarding dementia in Italian, Greek and Chinese Australians. Dement Geriatr Cognit Disord 2010;30(6):499–508.
[39] Mukadam N, Waugh A, Cooper C, Livingston G. What would encourage help-seeking for memory problems among UK-based South Asians? A qualitative study. BMJ Open 2015;5(9):e007990. https://doi.org/10.1136/bmjopen-2015-007990.
[40] Uppal G, Bonas S. Constructions of dementia in the South Asian community: a systematic literature review. Ment Health Relig Cult 2014;17(2):143–60.

[41] Chiarelli L, Edwards P. Building healthy public policy. Can J Public Health 2006;97(Suppl. 2):S37–42.
[42] de Wit L, Fenenga C, Giammarchi C, di Furia L, Hutter I, de Winter A, et al. Community-based initiatives improving critical health literacy: a systematic review and meta-synthesis of qualitative evidence. BMC Public Health July 20, 2017;18:40.
[43] Keefe J. The impact of ethnicity on helping older relatives: findings from a sample of employed Canadians. Can J Aging 2000;19(3):317–42.
[44] Kreps G, Neuhauser L, Sparks L, Villagran MM. The power of community-based health communication interventions to promote cancer prevention and control for at-risk populations. Transl community-based health. Commun Interv Promote Cancer Prev Control Vulnerable Audiences 2008;71(3):315–8.
[45] Blair TR. "Community ambassadors" for South Asian elder immigrants: late-life acculturation and the roles of community health workers. Soc Sci Med 2012;75(10):1769–77.
[46] Boughtwood DL, Adams J, Shanley C, Santalucia Y, Kyriazopoulos H. Experiences and perceptions of culturally and linguistically diverse family carers of people with dementia. Am J Alzheimers Dis Other Demen 2011;26(4):290–7.
[47] Gilson L. Trust and the development of health care as a social institution. Soc Sci Med 2003;56(7):1453–68.
[48] Genoff MC, Zaballa A, Gany F, Gonzalez J, Ramirez J, Jewell ST, et al. Navigating language barriers: a systematic review of patient navigators' impact on cancer screening for limited English proficient patients. J Gen Intern Med 2016;31(4):426–34.
[49] Wells KJ, Valverde P, Ustjanauskas AE, Calhoun EA, Risendal BC. What are patient navigators doing, for whom, and where? A national survey evaluating the types of services provided by patient navigators. Patient Educ Couns 2018;101(2):285–94.
[50] Bradford A, Kunik ME, Schulz P, Williams SP, Singh H. Missed and delayed diagnosis of dementia in primary care: prevalence and contributing factors. Alzheimer Dis Assoc Disord 2009;23(4):306–14.
[51] Bebbington P, Meltzer H, Brugha T, Farrell M, Jenkins R, Ceresa C, et al. Unequal access and unmet need: neurotic disorders and the use of primary care services. Psychol Med 2000;30(06):1359–67.
[52] Schoenmakers B, Buntinx F, Delepeleire J. What is the role of the general practitioner towards the family caregiver of a community-dwelling demented relative? A systematic literature review. Scand J Prim Health Care 2009;27(1):31–40.
[53] Koch T, Iliffe S. Rapid appraisal of barriers to the diagnosis and management of patients with dementia in primary care: a systematic review. BMC Fam Pract 2010;11(1):52.
[54] Pimlott N. Preventive care: so many recommendations, so little time. Can Med Assoc J 2005;173(11):1345–6.
[55] Tullo ES, Lee RP, Robinson L, Allan L. Why is dementia different? Medical students' views about deceiving people with dementia. Aging Ment Health 2015;19(8):731–8.
[56] Services de santé pour la communauté franco-ontarienne. Feuille de route pour une meilleure accessibilité et une plus grande responsabilisation (Working group on health services in French). Toronto: Services de santé pour la communauté franco-ontarienne; 2005. Available from: http://www.ontla.on.ca/library/repository/mon/14000/261725.pdf.
[57] Hinton L, Franz C, Friend J. Pathways to dementia diagnosis: evidence for cross-ethnic differences. Alzheimer Dis Assoc Disord 2004;18(3):134–44.
[58] Armstrong K, Rose A, Peters N, Long JA, McMurphy S, Shea JA. Distrust of the health care system and self-reported health in the United States. J Gen Intern Med 2006;21(4):292–7.

[59] Fitzpatrick MA, Vangelisti A. Communications, relationships and jealth. In: Robinson P, Giles H, editors. The new handbook of language and social psychology. Chichester, England: John Wiley and Sons Ltd.; 2001. p. 506–22.
[60] Borson S, Scanlan JM, Watanabe J, Tu SP, Lessig M. Improving identification of cognitive impairment in primary care. Int J Geriatr Psychiatry 2006;21(4):349–55.
[61] Lang L, Clifford A, Wei L, Zhang D, Leung D, Augustine G, et al. Prevalence and determinants of undetected dementia in the community: a systematic literature review and a meta-analysis. BMJ Open February 1, 2017;7(2):e011146. https://doi.org/10.1136/bmjopen-2016-011146.
[62] Amjad H, Roth DL, Sheehan OC, Lyketsos CG, Wolff JL, Samus QM. Underdiagnosis of dementia: an observational study of patterns in diagnosis and awareness in U.S. older adults. J Gen Intern Med July 1, 2018;33(7):1131–8.
[63] Tsoi KKF, Chan JYC, Hirai HW, Wong SYS, Kwok TCY. Cognitive tests to detect dementia: a systematic review and meta-analysis. JAMA Intern Med September 1, 2015;175(9):1450–8.

Further reading

[1] Lubetkin EI, Lu W-H, Krebs P, Yeung H, Ostroff JS. Exploring primary care providers' interest in using patient navigators to assist in the delivery of tobacco cessation treatment to low income, ethnic/racial minority patients. J Community Health 2010;35(6):618–24.

CHAPTER 3

Driving cessation in people with dementia

Bonnie M. Dobbs
Department of Family Medicine, University of Alberta, Edmonton, AB, Canada

Introduction

Transportation mobility is a key component in the maintenance of independence and well-being of older adults [1–3], with driving the primary mode of transportation for a large majority of adults aged 65+ years [4–6]. Transitioning from driver to non-driver is a normative life event in that the majority of older adults eventually stop driving. Specifically, men are expected to live 7 years beyond their ability to drive, with women expected to live 10 years beyond their driving ability [7]. In Canada, the percentage of adults 65+ years holding a valid driver's license decreases from 80% for those aged 65–74 years to less than 20% for those aged 90 years and older [6], with similar trends evident elsewhere [8].

A number of factors are associated with driving cessation in the 65+ population. These factors include age (older), sex (female), marital status (widowed/divorced), living arrangements (living alone), income (low), place of residence (suburbs and rural), and the presence of one or more medical conditions [5,6,9–12]. Some people make the decision to stop driving for themselves, as in the following illustration from the Dementia Transitions Study, as described in Chapter 1 [13]:

> *Person with dementia: I knew I could no longer drive. I told myself I don't know how to drive, I can't drive anymore, and I immediately gave my car to her brother on the weekend.*
>
> *Interviewer: Just like that?*
>
> *Person with dementia: Well yes. I didn't have my license anymore; I went to give it in. When I told myself, no in this situation, I want to give in my license; because I did not want to take any chance of hitting someone.*

But probably for most, the decision to stop driving is fraught with varying experiences of worry, frustration, anger, confusion, injustice,

Evidence-informed Approaches for Managing Dementia Transitions
ISBN 978-0-12-817566-8
https://doi.org/10.1016/B978-0-12-817566-8.00003-6

obfuscation, and deception. Years after the local police, with the full consent of the family, declined to return a car lost for the second time in the supermarket car park, the father of one of our editors continued his early morning practice of going outside, first opening the yard gate, then opening the garage doors, to discover, astonished, an empty space. "Car's been stolen, Neil." (Drummond N, April 24, 2019, oral communication; unreferenced).

Medical conditions and driving

A substantial body of literature now exists on the role that medical conditions play in impaired driving performance [14—19], with the first publication on medical conditions and driving published in 1965 [20]. One of the most elegant studies to date on medical conditions and driving was conducted by Diller et al. [15]. They compared the at-fault crash rates of drivers with different medical conditions in the state of Utah to age-, gender-, and place of residence-matched controls. In terms of single medical conditions, drivers with diabetes and other metabolic conditions, impairments in visual acuity, musculoskeletal abnormalities (e.g., osteoporosis), psychiatric conditions (e.g., schizophrenia, major depressive disorder, bipolar disorder), and those with epilepsy and other episodic conditions (e.g., syncope, narcolepsy, hypoglycemia, episodic vertigo) had an increased at-fault crash risk that was 1.5 to 2 times higher than drivers without those conditions [15,16]. Drivers with neurological conditions such as stroke, Parkinson's disease, or head injuries had at-fault crash rates that were more than double those of matched controls. Drivers with learning, memory, and communication impairments (e.g., dementia) had more than a three-fold increase in at-fault crash rates than matched controls, with this rate exceeding that of alcohol-related crashes [15,16] (Fig. 3.1). Recent data indicate that the increased risk of crashes and life lost in the older population in general continues with a notable increase in fatal crash rates, based on per vehicle miles travelled, starting at ages 70—74 years, with the highest rates for drivers 85 years and older [21].

Driving and dementia

There now is a significant body of literature on driving and dementia, with research indicating that dementia is a significant risk factor for driving impairment [22—28], with driving competence affected both by stage of

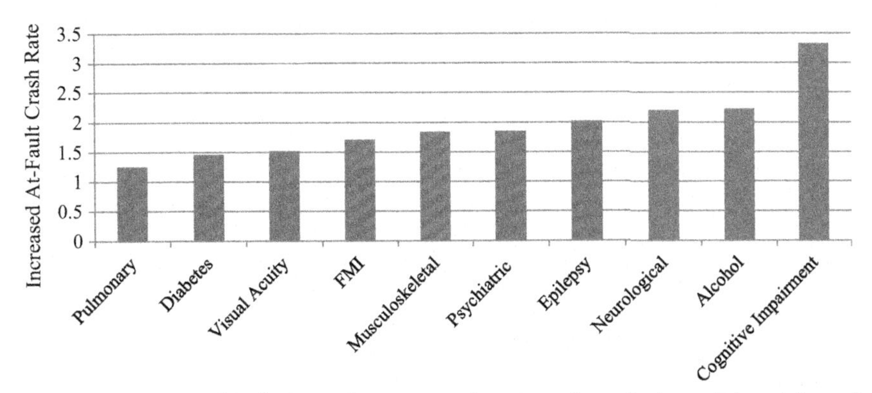

Figure 3.1 Increased at-fault crash rates as a function of medical condition. Adapted from "Evaluating Drivers Licensed with Medical Conditions in Utah, 1992–1996 (Report No. DOT HS 809 023)," by E. Diller, L. Cook, D. Leonard, J. M. Dean, J. Reading, and D. Vernon, 1999, Washington, DC: U.S. Department of Transportation. (Used with Permission).

the illness and pattern and degree of functional impairment [29–33]. As such, a diagnosis of dementia alone is not adequate enough to revoke an individual's license to drive [34]. The medical community plays a key role in determining medical "fitness to drive". To assist the medical community in this process, national guidelines and standards and legislation on reporting of medically unfit drivers are in place in many jurisdictions worldwide. A high-level overview of the national guidelines and standards, as well as legislation on reporting medically unfit drivers, is provided in the following section.

Determining medical "fitness to drive"

National guidelines/standards

In many jurisdictions, decisions regarding a person's medical "fitness to drive" are based on national guidelines or standards. Examples include the Canadian Council of Motor Transport Administrators (CCMTA) *Medical Standards Drivers* [35], *Assessing Fitness to Drive Medical Standards for Licensing and Clinical Management Guidelines* [36], *Medical Aspects of Fitness to Drive* [37], and *Mobility and Transport Road Safety Fitness to Drive* [38]. Licensing and "fitness to drive" decisions, however, are the responsibility of licensing authorities at the provincial/territorial/state level in many jurisdictions worldwide, with the national standards used by provincial, territorial, and/or state governments to establish whether drivers (both private and commercial) are medically "fit to drive". Also, there are national "fitness to drive" guidelines developed by medical associations to assist physicians and allied health professionals in the evaluation of medical "fitness to drive". Examples of these guidelines are the Canadian Medical Association's *Driver's Guide (9th Edition)* [39], the American Geriatrics Society's *Clinician's Guide to Assessing and Counseling Older Drivers (3rd Edition)* [40], and the Driver & Vehicle Licensing Agency of the United Kingdom's 2018 *Assessing Fitness to Drive: A Guide for Medical Professionals* [41].

Medical fitness to drive legislation

An overview of the legislation/regulations governing reporting of medically at-risk/impaired drivers in each jurisdiction worldwide is beyond the scope of this chapter. However, an overview of the legislation governing reporting of medically at-risk/impaired drivers in each of the Canadian provinces and territories, protection for reporting, and the admissibility of reports as evidence in legal proceedings is provided in Table 3.1 [39].

Table 3.1 Mandatory reporting by physicians legislation, protection for reporting, and physician–patient privilege in Canada by province/territory.

Province/Territory	Reporting	Physician[a] protection for reporting	Admissibility of reports as evidence in legal proceedings[b]
Alberta	Not directly addressed but interpreted as discretionary	Protected	The identity of the reporting physician remains confidential if the report was made in good faith
British Columbia	Mandatory for physician if unfit driver has been warned of the danger and still continues to drive[c]	Protected unless physician acts falsely or maliciously	Not addressed
Manitoba	Mandatory	Protected	Privileged[d]
New Brunswick	Mandatory	Protected as long as physician acts in good faith	Not addressed

Table 3.1 Mandatory reporting by physicians legislation, protection for reporting, and physician–patient privilege in Canada by province/territory.—cont'd

Province/ Territory	Reporting	Physician[a] protection for reporting	Admissibility of reports as evidence in legal proceedings[b]
Newfoundland and Labrador	Mandatory	Protected	Privileged[e]
Northwest Territories	Mandatory	Protected unless physician acts maliciously or without reasonable grounds	Privileged[f]
Nova Scotia	Discretionary	Protected	Not addressed
Nunavut	Mandatory	Protected unless physician acts maliciously or without reasonable grounds	Privileged[f]
Ontario	Mandatory	Protected	Privileged[e]
Prince Edward Island	Mandatory	Protected	Privileged[e]
Quebec	Discretionary	Protected	Privileged[g]
Saskatchewan	Mandatory	Protected as long as the physician acts in good faith	Privileged[e]
Yukon	Mandatory	Protected	Not addressed

[a]Used with permission of the Canadian Medical Protective Association (CMPA).
[b]Information in this column is subject to the access-to-information legislation of the respective province or territory.
[c]Pending legislation in British Columbia will change the province from a mandatory reporting province to a hybrid mandatory/discretionary reporting province.
[d]Not admissible as evidence for any purpose.
[e]Not admissible in evidence at trial except to prove compliance with reporting obligations.
[f]Not admissible in evidence or open to public inspection except to prove compliance with reporting obligations and in a prosecution of a contravention of section 330 … The person who is the subject of the report is entitled to a copy of the report upon payment of a prescribed fee.
[g]Not admissible in evidence except in cases of judicial review of certain decisions of the motor vehicle licensing authority.
Source: Canadian Medical Association. CMA driver's guide: determining medical fitness to operate motor vehicles. 9th ed. [Internet]. Ottawa (ON): Joule, Inc; 2017 [cited 2019 Jan 8]. Available from: https://joulecma.ca/evidence/CMA-drivers-guide.

As shown in Table 3.1, all provinces and territories in Canada impose a statutory duty on physicians (and other healthcare professionals [HCPs] in a number of jurisdictions) to report any patient who, in their opinion, has a medical condition that may impact their "fitness to drive". This duty is mandatory in seven Canadian provinces and all three territories. In three of the provinces (Alberta, Nova Scotia and Québec), the duty to report is discretionary, meaning that physicians/other HCPs can report but are not mandated to do so. All ten provinces and the three territories provide protection to physicians for reporting (but see Table 3.1 for caveats regarding protection in certain provinces). With respect to liability, physicians have been found liable for failing to report a medically impaired driver, notably in those provinces and territories with mandatory requirements [39].

An important consideration in physician reporting of medically at-risk/impaired drivers is the presence of fitness to drive legislation and regulations governing the reporting of medically at-risk/impaired drivers, physician awareness/knowledge of reporting legislation, as well as physician willingness to report. In general, physician awareness and knowledge of the reporting legislation are poor [42–45]. For example, Bruchbacher et al. [42] found that the majority (78%) of physicians reported little knowledge or training on the laws in British Columbia on determining driver fitness or reporting medically unfit drivers. Knowledge of the reporting legislation and reporting practices among physicians also differs as a function of legislation. In general, results indicate that physicians from mandatory reporting jurisdictions are more correctly aware of the legislation and are more likely to report medically unsafe drivers to the licensing authority than are physicians from jurisdictions with discretionary reporting [46,47]. Area of practice also affects reporting, with specialists more likely to be aware of and to report medically unfit drivers than general practitioners [48].

The introduction of mandatory reporting legislation facilitates physician reporting of medically unfit drivers [45,47,49]. Medical warnings (reporting patients with medical conditions who may be unsafe to drive with the inclusion of a billing fee for the associated physician's services) also have been shown to facilitate reporting of patients who are unfit to drive and have been effective in reducing subsequent risk of road crashes [50]. Implementation of educational programs, along with publication of national guidelines on medical fitness to drive, have been shown to be associated with a high degree of family physician awareness of this topic [51,52].

Identified barriers to physician reporting of medically impaired drivers to the licensing authority include lack of time, lack of knowledge of guidelines for reporting or of the process for doing so, lack of remuneration, unawareness of the legal requirement for reporting, and breach of confidentiality [42,51]. The negative impact that reporting has on the physician—patient relationship, the fear of losing patients or the experience of their actual loss, and pressure from patients or families not to report also have been identified as barriers to physician reporting [42,50]. However, as noted by Weir [53], physicians are obligated to intervene when the patient's medical condition presents harm to others and/or to self.

Voluntary reporting of medically unfit drivers

In many jurisdictions, drivers who may be at-risk to the public can be reported to the licensing authority by concerned stakeholders (e.g., family, law enforcement, physicians, social service professionals) [54] or the public at large [55]. Information related to the complaint remains confidential, with the licensing authority taking action only when there are reasonable grounds to believe that the reported driver is a safety risk to him/herself or to the public [55]. In a recent study, Meuser et al. [56] found that 65% of the drivers reported to the driver licensing agency by family members had some cognitive issue (e.g., confusion, memory loss, becoming lost while driving), with few of the family-reported drivers (2%) retaining a valid driver's license. In addition, the written concerns of family members were validated, for the most part, by subsequent physician evaluation and license examination. These results suggest that family members are important stakeholders with respect to both driver and public safety.

The introduction of a voluntary reporting law, which allows concerned physicians, police officers, driver licensing office staff, family members, and others to report potentially at-risk drivers for re-evaluation and possible license revocation if they are found to be medically unfit to drive has been found to be effective in preventing drivers who are unsafe from continuing to drive [56]. Specifically, in a study that investigated the effectiveness of a voluntary reporting law, Meuser et al. [45] found that the crash involvement of reported drivers decreased rapidly after introduction of the voluntary reporting law, with a fourfold reduction in crashes over the 5-year reporting period. With respect to sources of reporting medically impaired drivers to the licensing authority, Meuser et al. [56] found that reports were submitted by police officers (30%), driver licensing office staff (27%), physicians

(20%), family members (16%), and others (7%), with the most common reported medical condition being dementia/cognitive impairment (45%). Finally, some jurisdictions require drivers to "self-report" medical conditions (e.g., loss of consciousness, vision impairments) that may trigger licensing actions. However, a lack of awareness of self-reporting requirements and poor compliance when the driver is aware limit the impact of this policy [57].

Driving and dementia

Identification and assessment of "fitness to drive" in people with dementia

Driving in people with dementia is an important public health concern in that all drivers with dementia will become unsafe to drive at some stage in their illness. Testing fitness to drive can be traumatic for all concerned, as demonstrated in these statements by participants in the Dementia Transitions Study:

> *Person with dementia: Well he was picking holes ... now look what you done here; now look what you done here. Now what do you expect, I answered the questions to the best of my ability ... Anyway, I mean, memory has nothing to do with driving. You know what I mean, nothing to do. [ID 3023]*

> *Daughter of a person with dementia: That was when he lost his license. That was it. He could not understand that. It didn't matter how much I talked to him, or how much Dr. X talked to him. He was mad at Dr. X ... and he must have forgotten that it was Dr. X that did the first test. He said to Dr. X: "I went to see somebody here, and some bloody idiot took my license away." [ID 1032]*

In Canada and many other countries, family physicians are generally the patients' first point of contact with the health system and, as such, are "ideally positioned to provide care to individuals living with dementia from early to end stages of the illness" [58] [pp. 717]. To guide family physicians in that process, consensus statements on the diagnosis and treatment of dementia and medical "fitness to drive" guidelines have been published [59,60]. As indicated in the study by Moore [61] [pp. 433—434], successive "Canadian Consensus Conference on the Diagnosis and Treatment of Dementia (CCCDTD) statements recommended that diagnosis and management of patients with dementia should mainly be the responsibility of primary health care and this remains implicitly valid." There is support for this recommendation from primary care physicians in that results from a recent national survey indicate that more than 86% of Canadian primary care physicians "often" or "sometimes" provide medical care for persons with dementia [62]. In addition to diagnosing dementia, communicating the diagnosis to persons with dementia and their family, and providing post-diagnosis treatment, support, and advocacy [58], family physicians also play an important role in identifying persons with dementia who may no longer be safe to drive [58,63,64]. There is mixed evidence with respect to the comfort of family physicians' in diagnosing and managing persons with dementia with some studies indicating that the majority of family physicians are comfortable with diagnosing and managing persons with dementia, with about half of family physicians comfortable in dealing with driver's license issues [65,66]. However, other studies indicate a lack of comfort and/or confidence in all three of these areas [51,64,67].

An important consideration in medical "fitness to drive" is the *impact* of the condition and/or conditions on the *functional ability to drive* as opposed to the *presence* of the medical condition and/or conditions [14]. A number of in-office screening tools have been developed to assist physicians in identifying patients who are cognitively impaired. Common in-office screening tests include the Mini-Mental State Examination (MMSE) [68], the Montreal Cognitive Assessment (MoCA) [69], the Trail-Making Tests (TMT) A and B [70], and the clock drawing test [71]. Although these screening tools were not designed to identify patients who may no longer be "fit to drive," they are commonly used for that purpose. However, there is an emerging body of research that suggests that the effectiveness of these tests for predicting driving competence is questionable [68–72]. To address this gap, Dobbs and Schopflocher [73] developed the **S**creen for the **I**dentification of **M**edically-**A**t-**R**isk **D**rivers A **M**odification of the **D**emtect (SIMARD MD) to assist healthcare practitioners in identifying drivers with dementia who may no longer be safe to drive. The difference between common in-office screening tools for identifying cognitive impairment and the SIMARD MD is that the former tests were developed to identify cognitive impairment and not driving competence, whereas the SIMARD MD was developed specifically to identify driving competence in drivers with cognitive impairment/dementia [73]. Scores on the SIMARD MD range from 0 to 130, with established cut points to identify drivers who have a high probability of failing an on-road driving assessment, those who have a high probability of passing an on-road driving assessment, and those in the indeterminate range who are in need of an on-road assessment to determine driving competency. The paper and pencil test is easy to administer and can be scored in one to two minutes with no special training or clinical expertise needed. The SIMARD MD is available at no cost to physicians and allied healthcare professionals, with access to the screening tool via a password-protected website (see mard.ualberta.ca).

In addition to in-office screening tools, recent research suggests that cerebrospinal fluid (CSF) biomarkers for Alzheimer's disease (AD) predict marginal or fail ratings on an on-road driving assessment [74], with the authors concluding that driving performance shows promise as a functional outcome in AD prevention trials [74]. Stout et al. [75] also found that high levels of CSF biomarkers for AD were predictive of a shorter time to driving cessation, with persons with dementia with abnormal CSF biomarker levels having stopped driving at about twice the rate per year as compared with those with normal CSF biomarker levels.

Finally, caregivers of persons with dementia often are involved in decisions related to driving, with physicians often relying on the caregiver's report of the patient's ability to drive [76]. However, results from the published literature indicate that both patients and caregivers often overestimate the person with dementia's competency to drive [77–80].

Given that some patients with dementia demonstrate decreased awareness of deficits [77] and a strong motivation to continue driving, collaterals are frequently asked about the patient's history of accidents, tendency to become lost, and general driving ability. On the basis of the literature reviewed, caregivers' reports of driving ability do not correlate with patients' memory, executive, or language functioning. Rather, only changes in mental status and visuospatial skills indicate that caregivers are more likely to report reduced driving skills [28].

On-road testing is considered the gold standard for determining "fitness to drive" in persons with dementia [81–83]. However, on-road assessment often is not available in rural and remote areas, which presents challenges for physicians in their assessment of "fitness to drive" in this patient population. In general, most persons with dementia in the *mild* stages of the illness will be fit to drive, with a recommendation for regular off- and on-road reassessment of driving competency every 6–12 months [39]. However, as the disease progresses, there is evidence and general agreement that persons with dementia in the moderate to severe stages of the illness should not drive [39,84]. One of the challenges for healthcare professionals (and for persons with dementia and their family) is in determining the most appropriate time to cease driving.

Physician discussions related to "fitness to drive" in dementia

The loss of driving privileges due to a dementia impacts a sizable number of persons with dementia today and will continue to do so over the next several decades due to the aging of the baby boomers [85]. For many persons with dementia, the loss of driving privileges is a major occurrence in the course of their illness, with the negativity associated with this loss equated by some to the impact associated with receiving the diagnosis [86–88]. As noted earlier, discussions related to driving are difficult and one of the most challenging aspects that physicians (and family members) face in caring for persons with dementia [88,89]. Commonly identified practice stressors related to fitness to drive include differences of opinion about competency to drive among patients, families, and the physician [64,90]; inadequate physician training [91,92]; fear of jeopardizing the long-standing, patient–physician relationships [93,94]; and time constraints [95]. Referral to a specialist for issues related to driving can help family physicians in preserving the physician–patient relationship [64].

Initiating discussions about driving in the early stages of the illness when the person with dementia still has insight and can appreciate the need for stopping driving can help the person with dementia and his/her family members prepare for the day that the person with dementia has to transition from the driver's seat to the passenger's seat [96]. Early discussions also allow for planning for alternate means of staying mobile when driving is no longer an option [30]. However, conversations on the need for driving cessation often are very difficult for the person with dementia and their family in that driving is not only a primary means of meeting one's transportation needs, it also is a symbol of independence and autonomy for the person with dementia [96,97]. A number of resources are available to assist physicians in dealing with the driving issue. The 9th edition of the Canadian Medical Association *Driver's Guide* [39] has an excellent section on strategies for physicians for discussing driving cessation in persons with dementia (Table 3.2).

Dementia and driving cessation

Driving cessation is inevitable for all drivers with a dementia due to the progressive decline in cognitive abilities [84,98]. However, the challenge for the person with dementia, his/her family, and healthcare professionals is in determining the most appropriate time to cease driving.

Table 3.2 Strategies for discussing driving cessation in dementia.

- Before the appointment, consider the patient's impairments. It may be important to ask if the spouse or another caregiver can be present. This can provide emotional support and help to ensure that the family understands that the person needs to stop driving. It also may be helpful to meet with the family (with the patient's consent) before holding a meeting with the patient and family.
- Whenever possible, the appointment should be in a private setting where everyone can be seated. Always address the patient preferentially, both in the initial greeting and in the discussion.
- For patients with progressive illnesses, such as dementia, discuss driving early in the course of the condition, before it becomes a problem. Early discussions also allow patients and family members to prepare for the day when driving is no longer an option.
- Be aware that patient and caregiver reports of driving competence often do not reflect actual competence. Evidence of impaired driving performance from an external source (e.g., driving assessment, record of motor vehicle crashes or near misses) can be helpful. Discuss the risks of continuing to drive with the patient and family members.
- For patients who are medically unfit to drive, emphasize the need to stop driving using the driving assessment, if available, as the appropriate focus.
- Often the patient will talk about his or her past good driving record. Acknowledge that accomplishment in a genuine manner, but return to the need to stop driving. Sometimes saying "Medical conditions can make even the best drivers unsafe" can help to refocus the discussion.
- It is common for drivers, especially those who are older, to talk about a wide range of accomplishments that are intended, somehow, to show that there could not be a problem now. Again, acknowledge those accomplishments, but follow with "Things change. Let's not talk about the past. We need to focus on the present." to end that line of conversation and refocus the discussion.
- Ask how the person is feeling and acknowledge his or her emotions. Avoid lengthy attempts to convince the person through rational explanations. Rational arguments are likely to evoke rebuttals.
- It is likely that emotions and feelings of diminished self-worth represent a real issue behind resistance to accept advice or direction to stop driving. Explore these feelings with empathy. A focus on the feelings can deflect arguments about the evaluation and the stop-driving directive.
- Ask the patient what he or she understands from the discussion. It may be important to schedule a second appointment to further discuss the patient's response and explore next steps.
- Document all discussions about driving in the patient's chart.

Table 3.2 Strategies for discussing driving cessation in dementia.—cont'd

• To assist patients in staying mobile, have them create a "mobility account" using the money that they would have used to own and operate their own vehicle. For example, it is estimated that for vehicles driven about 24,000 km a year, average car ownership costs were $8469 a year or about $706 per month. The purpose of the mobility account is to have funds set aside to cover the costs of alternative transportation.

Reproduced, with updates, with permission from Canadian Medical Association. CMA driver's guide: determining medical fitness to operate motor vehicles. 9th ed. [Internet]. Ottawa, ON: Joule, Inc; 2017 [cited 2019 Jan 8]. Available from: https://joulecma.ca/evidence/CMA-drivers-guide.

Factors associated with driving cessation in persons with dementia—intrapersonal, interpersonal, and environmental

A number of factors are associated with driving cessation: intrapersonal, interpersonal, and environmental [99]. Intrapersonal factors include age, gender, presence of an illness that affects functional abilities to drive, awareness of the decline in physical, visual, and cognitive abilities, as well the individual's opinion of the importance of driving and one's own driving safety. Interpersonal factors include the family member's, and authority figure's, opinions regarding driving safety. Environmental factors include crash risk and availability of alternate forms of transportation.

With respect to intrapersonal factors, the type of dementia affects the likelihood of driving cessation. Specifically, drivers with AD are more likely to continue to drive after diagnosis than individuals diagnosed with other types of dementia [100—102]. Seiler et al. [103] found that a higher percent of drivers with Lewy body dementia (DLB) ceased driving, with a rate of 91% as compared with 55%—67% of drivers with AD, vascular dementia (VD), and frontotemporal dementia (FTD) having ceased driving. Stage of illness also is a factor, with an estimated one in three persons with mild to moderate dementia continuing to drive [104—106]. Of interest, there now is evidence that deterioration in cognitive performance begins *9 to 12 years before* diagnostic criteria for dementia are met, with the cognitive deficits deemed sufficient enough to affect driving safety occurring *a few years* before diagnostic criteria for dementia are met [107—110]. Consistent with this evidence, Marie Dit Asse et al. [110] have found that some persons with dementia stop driving approximately 2 years before a diagnosis of dementia, with women more likely to stop before diagnosis than their male counterparts. However, rates of stopping driving approximately 3 years after diagnosis are similar for males and females [110]. Early-onset dementia is associated with a higher probability of driving cessation [111]. Other intrapersonal factors related to dementia and driving cessation include older age [105,112], female gender [103,113, but see 80,114], and place of residence (urban-dwelling) [105,110]. Lack of awareness of the decline in cognitive abilities (anosognosia) also is an important factor in terms of driving with persons with dementia. Although anosognosia is present in a wide range of neurological disorders, it is particularly common in dementias [115], with anosognosia identified in approximately 40% of persons with dementia [116,117]. Anosognosia can lead to persons with dementia overestimating their driving abilities [118–120], as well as being less likely to regulate their driving behaviour [121].

Interpersonal factors associated with driving cessation in persons with dementia include the family members' and authority figures' opinions regarding driving safety. The risk estimate of caregivers has been found to be a significant predictor of driving cessation in persons with dementia [103]. However, studies show that approximately half of family members who express doubt about a family member's lack of driving competence do not attempt to engage in efforts related to driving cessation [122]. In addition, family members who rely on the person with dementia for transportation encourage continued driving despite evidence that the person with dementia is no longer fit to drive [123,124]. On the other hand, recommendations for driving cessation from authority figures (e.g., physicians) facilitate driving cessation [122,123].

Environmental factors are the last set of factors that affect driving cessation in persons with dementia, with motor vehicle crashes and revocation of the driver's license associated with driving cessation in this patient population [103,124]. The availability of alternate transportation has been found to facilitate driving cessation in persons with dementia, with persons with dementia living in urban settings where alternate transportation is available more likely to stop driving [105]. However, the majority of person with dementia depend on family members and/or friends for their transportation mobility [125].

Stages of driving cessation

Three stages of driving cessation have been identified in persons with dementia: the *worried waiting stage*, the *crisis stage*, and the *postcessation stage* [126]. The *worried waiting stage* involves balancing safety with impending losses and challenges of knowing when to stop. The *crisis stage* involves risky driving or difficult transportation, acute adjustment to cessation and to life without driving, and relationship conflict. The *postcessation stage* is described as a long journey with ongoing battles and adjustments and decreased life space, with this stage affected by disease progression and exhaustion of the caregiver [126]. Driving cessation in persons with dementia can be voluntary or involuntary. Although some persons with dementia cease driving voluntarily, the transition from the driver's seat to the passenger seat is involuntary for many persons with dementia [32,102,127–129].

Supporting the person with dementia and family through the loss of driving privileges

Most everyone prepares for retirement from work, yet few of us prepare for retirement from driving. As such, most persons with dementia and their family will need to start preparing for the day when the person with dementia transitions from the driver's seat to the passenger seat. Strategies to help in this transition include creating an "advanced driving directive" at the time of advance care planning to let the person with dementia's family physician and/or specialist, as well as family members know of their wishes for their driving and mobility future. In an examination of older adult's opinions on advanced driving directives, the majority of adults 55 years and older were open to driving discussions, with 54% of participants indicating that they would be willing to complete an advanced driving directive if recommended by a physician, friend, or family member, with another one-third indicating "maybe" [130]. In addition, of those indicating yes, 79% said it was "likely" or "very likely" they would comply with the advanced driving directive in the future [130]. An additional advantage of an advanced driving directive is that it can facilitate the difficult discussion when it comes time for revocation of the license [130]. Agreements to "stop driving" also can be helpful (see Fig. 3.2 for an "Agreement with Family") [131].

For family members, it is important to recognize that bringing up the driving topic is almost always difficult. Focusing on the impact of the illness rather than on past driving abilities can help to keep the discussion focused on the need to stop driving [30]. Examples of conversations starters are provided below [132].

"I know you've been a good driver for a long time, but things have changed ..."

"Dad, I'm really concerned about your driving – you have to stop now before something serious happens ..."

"Mom, you have been such a good driver for so long, let's not let it end with something terrible happening ..."

AGREEMENT WITH MY FAMILY ABOUT DRIVING

To My Family:

The time may come when I can no longer make the best decisions for the safety of others and myself. Therefore, to help my family member make necessary decisions, this statement is an expression of my wishes and directions while I am still able to make these decisions.

I have discussed with my family member my desire to drive as long as it is safe for me to do so.

When it is not reasonable for me to drive, I desire ____________________ (person's name) to tell me I can no longer drive.

I trust my family member will take the necessary steps to prohibit my driving to ensure my safety and the safety of others while protecting my dignity.

Signed: ____________________ Date: ____________________

Copies of this request have been shared with:

____________________ ____________________

____________________ ____________________

____________________ ____________________

____________________ ____________________

Figure 3.2 Agreement with My Family About Driving. Source: "At the Crossroads Guidebook", The Harford: © 2019 by The Hartford, All rights reserved. (Used with permission). https://www.thehartford.com/resources/mature-market-excellence/publications-on-aging.

Lived experiences in transitioning from driver to non-driver

There are few studies that have examined the person with dementia's lived experiences of transitioning from the driver's seat to the passenger seat. Using qualitative methodology, Seguin [133] found that persons with dementia might not understand the link between cognitive assessments, their memory, and their driving performance. For a small number of persons with dementia who understand why they are no longer allowed to drive, they seem to accept and self-regulate their driving cessation.

It is also the case that driving meets different needs for older adults. Those needs include accessing essential services (e.g., medical appointments, shopping), social/recreational, and esthetic purposes (e.g., a drive to the park or countryside). As noted by Musselwhite and Haddad [134] [pp. 31], "[Q]uality of life is enhanced by the very fact that the journey is not necessary. They often view quality of life as immersion in the world beyond everyday survival and completion of ordinary tasks …".

Results from Driving Cessation Support Groups (DSCGs) at the University of Alberta [135] support this finding. When asked "what things you no longer do because you are unable to drive," one of the male participants indicated that *"I don't go shopping in hardware stores because the bus routes are difficult."* [Participant 3], with a female participant indicating that *"The biggest thing that I am unable to do since losing my license is getting to and from my exercise classes and [I've] been unable to get to the library." [Participant 4].* Another participant indicated that she *"no longer goes shopping, or … to church … or to visit friends. All parts of my life are affected because I am not able to drive" [Participant 7].*

The loss of driving in persons with dementia affects not only the person with dementia but also the spouse, their children, as well as relatives and/or friends. Results from a recent study on the impact of driving cessation on family members (spouses, long-term partners, children, siblings) indicated that most supporters were negatively impacted [136]. Common themes included intrusion into the personal space of the supporter, as well as resentment and frustration, with an associated increase in stress. For example, one female supporter indicated that *"I don't seem to have any time to myself. [He] is here a lot and I don't seem to get a lot of space … There is no escaping. Personal space is important … It's like punishment in a way."* [136] [pp. 498].

A number of strategies that caregivers can use to prevent the person with dementia from driving include removing the keys, disabling the vehicle, or selling the vehicle. Resources, such as the *Conversations About Driving and Dementia* [137] are available to assist persons with dementia, family members, and healthcare professionals in having conversations on sensitive topics such as driving cessation (see https://alzheimer.ca/sites/default/files/files/national/brochures-conversations/conversations-about-dementia-and-driving.pdf). Making plans to help the person with dementia (and often the spouse) stay mobile also is important in the early stages of the illness.

Driving cessation support groups

Losing a license can be a stressful and emotionally devastating event for both the person with dementia and their family members/caregivers. Despite the recognition of the impact and the salience of the loss of driving privileges for persons with dementia and their family members/caregivers, there are few programs available directly designed to assist individuals and their caregivers in coping with this negative event [138]. The majority of support programs that are available are designed for caregivers of persons with dementia, with the focus on educational information about the disease, changes in lifestyles and abilities, expression of feelings, and practical advice [139–141], with the results showing significant positive effects on the caregivers' psychological well-being [140]. However, few support programs are available to assist persons with dementia and their families cope with the loss of a driver's license. To address this need, Driving Cessation Support Groups (DCSGs) were developed by Dobbs et al. in the early 2000s to assist persons with dementia and their families cope with the loss of driving privileges. The DCSGs are based on a structured protocol [142], with the person with dementia group led by a clinical psychologist (or by an individual with extensive therapeutic experience with older adults) and the caregiver group led by a social worker or other healthcare professional who has been trained to lead the groups. The DCSGs are held weekly, with the person with dementia and their caregivers meeting concurrently and independently over a 16-week period. There is no cost for attending the groups and transportation is provided at no charge. Over the past 20 years, 12 waves of DCSGs have been held in Edmonton and Area.

Efficacy of the groups for persons with dementia is evaluated through objective measures of depression, quality of life, behavioural disturbances, and through the person with dementia's perceptions of coping with loss of driving privileges, with comparison in measures for the DCSG participants to control group participants [143]. Efficacy of the groups for caregivers is evaluated through objective measures of depression, caregiver burden, caregiver competency, and coping self-efficacy [143]. Outcomes from both the persons with dementia and the caregiver groups have been consistent across DCSGs in the past 2 decades (see Figs. 3.3 and 3.4 for examples of results).

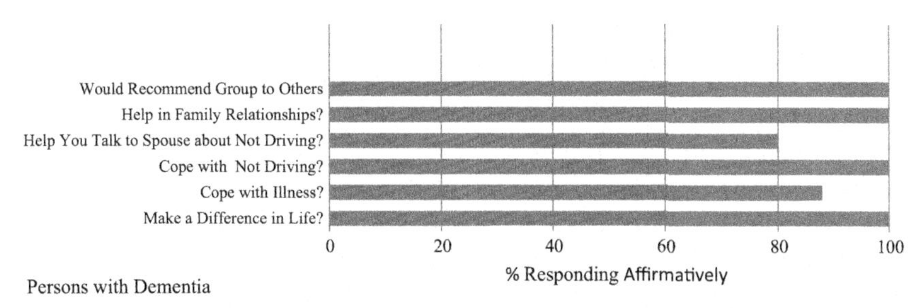

Figure 3.3 Outcomes from DCSGs – PwD Group. Dobbs B, Harper L, Pisani L, Visram F, Bhardwaj P. (2016). *Implementation of Driving Cessation Support Groups for Individuals with Dementia and their Caregivers in Covenant Health*. Presented at the Covenant Health Research Day, February 7, 2017, Edmonton, Alberta, Canada.

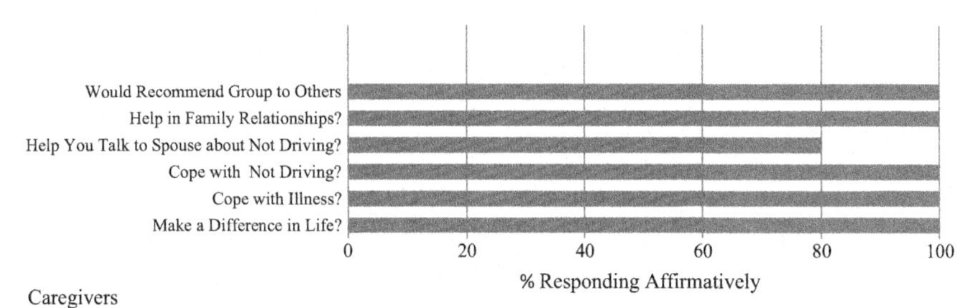

Figure 3.4 Outcomes from DCSGs – Caregiver Group. Dobbs B, Harper L., Pisani L., Visram F, Bhardwaj P. (2016). *Implementation of Driving Cessation Support Groups for Individuals with Dementia and their Caregivers in Covenant Health*. Presented at the Covenant Health Research Day, February 7, 2017, Edmonton, Alberta, Canada.

In addition, although the DCSGs were designed primarily to assist with the negative consequences associated with loss of driving privileges, results indicate that the DCSGs also are beneficial in helping the persons with dementia cope with their illness and in talking with family members, including the spouse, about the driving issue [143]. Group interventions have been developed in other jurisdictions to assist persons with dementia in coping with the loss of a driver's license. Examples of those group interventions include the UQDRIVE program in Australia [126], the *Traveling a New Road* [138], and *At the Crossroads* [141].

Enhancing mobility after driving cessation in persons with dementia

Enhancing the mobility of persons with dementia following driving cessation is an important consideration in that a number of persons with dementia cease driving in the early stages of their dementia. Although public transit and taxis are available in all large and small urban centers in Canada, only a very small minority of adults aged 65 years and older use either of these types of transportation as their primary means of transportation [6]. And, outside of the large and small urban centers, the alternatives to the car are virtually nonexistent as primary means of travel, with only 1% of seniors living outside these centers having reported that their primary form of transportation was accessible transit or taxis [6]. Barriers to use of public transit for persons with dementia include getting lost or confused to and from their destination, forgetting their intended destination, difficulty managing transactions, and/or difficulty coping with unexpected situations [144,145].

Alternate transportation services (e.g., senior buses, handivans, volunteer driving programs) outside of public transit can enable the safety and mobility of persons with dementia (and their caregivers) who have stopped driving [146]. However, there often are significant barriers to use of these types of services not only for older adults in general, but particularly for persons with dementia. Those barriers include knowledge of and/or hours of service availability, distance to destination, eligibility requirements, and the need for an accompanying companion [147]. Strategies to overcome these barriers include marketing and outreach to increase awareness of services, driver training on cognitive impairment and dementia, and the development and implementation of policies that enable the safety and mobility of persons with dementia (e.g., complimentary rides for an accompanying family member or friend, door-through-door service). An important consideration with respect to alternate transportation services in both urban and rural communities is the sustainability of these types of services. The absence of dedicated streams of funding is one of the biggest challenges faced by alternate transportation service providers [147,148]. Finally, the use of ridesharing services such as Uber and Lyft holds promise for enhancing the mobility of seniors [149], with both of these ridesharing services having developed programs specifically for older adults and persons with disabilities (Lyft Concierge and Uber in partner with Circulation) (see https://www.healthcareitnews.com/news/uber-teams-circulation-transport-patients-doctors-appointments). Barriers to the use of the ridesharing services by older adults in general, and particularly for person with dementia, is the need for ownership of technologies to access these services (e.g., mobile devices/smart phones, and apps) [150], difficulty with access to the technology, managing the technology, technology anxiety, and trust, as well as their high cost [149]. To bridge this gap, third parties, such as

GoGoGrandparent, have created hotlines that the person with dementia or family members/friends can call, with a small fee added to the cost of each trip (see https://gogograndparent.com).

The continuing development of automated or self-driving cars is expected to enable the future mobility of transportation disadvantaged segments of our society. However, results from recent surveys indicate that the majority of older adults are not only reluctant to travel in a fully automated autonomous vehicle (AV), they have little interest in owning or leasing a vehicle that is highly automated [151,152]. Moreover, older adults hold neutral or negative opinions about the technology [151,152] and have the lowest propensity to adopt fully automated AVs [153]. In addition, even if there was a change in attitudes by older adults toward the use of fully automated AVs, barriers that keep many older adults from driving also preclude their use of AVs. Those barriers include difficulty entering and exiting the vehicle for those with mobility challenges, difficulty in getting to and from the vehicle from the home and the desired destination for those with physical and/or cognitive impairments, and communication between the occupant and the vehicle [154].

An important consideration for enhancing the transportation mobility of older adults, including persons with dementia, is the awareness of community support services. In general, older adults' awareness of community support services is low with the telephone directory, doctors' offices, and word of mouth seen as the most important sources of information on community support services [155]. In a follow-up study that focused on older adults' awareness of community health and support services for dementia care, Ploeg et al. [156] found that more than 37% of respondents identified their physician as the most important source of information on community health and support services, with 33% identifying family and neighbours and 31% identifying home health services as important information sources. In addition, knowing where to find information about community support services was associated with an increased likelihood of mentioning physicians and home health services as sources of assistance. In-office resources for physicians and allied healthcare professionals that can assist families in coping with loss of the license include the previously mentioned *Conversations About Dementia and Driving* [137], as well as *At the Crossroads: Family Conversations about Alzheimer's Disease, Dementia & Driving* [131] and *We Need to Talk Family Conversations With Older Drivers* [157]. All of these resources are available for download or for ordering at no cost (see https://www.thehartford.com/resources/mature-market-excellence/

publications-on-aging). However, as noted earlier, knowledge of community support services often is lacking, with access to those services difficult [131]. To ensure a better life for persons with dementia and their families, *dementia-friendly communities* are being established in many jurisdictions worldwide [158]. In addition to addressing the stigma and social isolation associated with dementia, the availability of suitable transportation options have been identified as an important attribute for a community to become dementia-friendly [159]. As noted by one care partner of a person with dementia in Saskatchewan:

> *"When it was obvious that my mother's dementia had reached a point where it was concerning, she moved to where care was more available and she was closer to family. But she lost her sense of community, her closest friends, and those most likely to understand and provide support." Anonymous quote from care partner surveyed in* Dementia-Friendly Communities: Stakeholder Consultation Report *(2017) [159].*

Linking persons with dementia and their caregivers to community support services, including information on alternate means of transportation, can help to enhance the safety and mobility of persons with dementia and often the caregiver. Examples of resources that enable independence and mobility for persons with dementia and their family members include *Guides to Mobility and Independence* [160,161] and an *Online Listing of Alternate Transportation for Seniors Services* [162] (see www.mard.ualberta.ca).

Conclusion

The aging of the population, along with the strong association between age and the prevalence of dementia, will result in a greater number of older adults who will have to transition from behind the wheel over the next several decades. Family physicians play an important role in determining "fitness to drive", with medical "fitness to drive" guidelines in place in many jurisdictions to assist physicians in the decision-making process. The loss of a driver's license is a major event in the lives of persons with dementia as well as their families. Understanding the intrapersonal, interpersonal, and environmental factors associated with driving cessation in persons with dementia can assist physicians, families, and friends in preparing for and supporting the person with dementia (and their family members) through the loss of driving privilege. Knowledge and use of alternate transportation services and other relevant community resources can assist persons with dementia and their caregivers in maintaining their safety and mobility, along with enhancing their quality of life.

References

[1] Carp FM. Significance of the mobility of the well-being of the elderly. In: Transportation Research Board, editor. Transportation in an aging society: improving mobility and safety for older persons — volume 1 and volume 2 — special report 218. Washington (DC): The National Academies Press; 1988. p. 1–20.

[2] Gillins L. Yielding to age: when the elderly can no longer drive. J Gerontol Nurs [Internet] November 1990;16(11):12–5. https://doi.org/10.3928/0098-9134-19901101-05 [cited 2019 Jan 7].

[3] Jeffery B, Muhajarine N, Johnson S, McIntosh T, Hamilton C, Novik N. An overview of healthy aging strategies in rural and urban Canada. [Internet]. Saskatchewan: Saskatchewan population health and evaluation research unit. University of Regina and University of Saskatchewan; 2018. Available from: http://spheru.ca/publications/files/Healthy Aging Enviro Scan Report June 2018 FINAL%2026-Sep-2018.pdf [cited 2019 Jan 7].

[4] Binette J, Vasold K. 2018 Home and community preferences: a national survey of adults age 18-plus [Internet]. Washington (DC): AARP Research; August 2018. Available from: https://www.aarp.org/research/topics/community/info-2018/2018-home-community-preference.html [cited 2019 Jan 7].

[5] Choi M, Adams KB, Kahana E. Self-regulatory driving behaviors: gender and transportation support effects. J Women Aging [Internet] April 2013;25(2):104–18. https://doi.org/10.1080/08952841.2012.720212 [cited 2019 Jan 7].

[6] Turcotte M. Profile of seniors' transportation habits. Catalogue no. 11-008-X. [Internet]. Ottawa (ON): Statistics Canada; January 2012. Available from: https://www150.statcan.gc.ca/n1/pub/11-008-x/2012001/article/11619-eng.htm [cited 2019 Jan 7].

[7] Foley D, Heimovitz H, Guralnik J, Brock D. Driving life expectancy of persons aged 70 years of age and older in the United States. Am J Public Health [Internet] August 2002;92(8):1284–9. https://doi.org/10.2105/AJPH.92.8.1284 [cited 2019 Jan 7].

[8] U.S. Department of Transportation. Federal Highway Administration. Licensed drivers by age and sex [Internet]. Washington (DC): U.S. Department of Transportation, Federal Highway Administration; 2018. Available from: https://www.fhwa.dot.gov/ohim/onh00/bar7.htm [updated 2018 Mar 28; cited 2019 Jan 7].

[9] Mackett R. The health implications of inequalities in travel. J Transp Health [Internet] September 2014;1(3):202–9. https://doi.org/10.1016/j.jth.2014.07.002 [cited 2019 Jan 7].

[10] Ryser L, Halseth G. Resolving mobility constraints impeding rural seniors' access to regionalized services. J Aging Soc Policy [Internet] June 2012;24(3):328–44. https://doi.org/10.1080/08959420.2012.683329 [cited 2019 Jan 7].

[11] Shergold I, Parkhurst G, Musselwhite C. Rural car dependence: an emerging barrier to community activity for older people. Transport Plan Techn [Internet] January 2012;35(1):69–85. https://doi.org/10.1080/03081060.2012.635417 [cited 2019 Jan 7].

[12] Statistics Canada. Living arrangements of seniors: families, households, and marital status; structural type of dwelling and collectives; 2011 census of population. Report No. 98-312-X2011003 [Internet]. Ottawa (ON): Statistics Canada. 2012. Available from: http://www12.statcan.gc.ca/census-recensement/2011/as-sa/98-312-x/98-312-x2011003_4-eng.pdf [cited 2019 Jan 7].

[13] Drummond N, McCleary L, Garcia L, McGilton K, Molnar F, Dalziel W, Jing Xu T, Turner D, Triscott J, Freiheit E. Assessing determinants of perceived quality in transitions for people with dementia: a prospective observational study. Can Geriatr J [Internet] March 2019;22(1):13–22. https://doi.org/10.5770/cgj.22.332 [cited 2019 May 7].

[14] Dobbs BM. Medical conditions and driving: a review of the literature (1960–2000). U.S DOT/NHTSA Publication No. DOT HS 809 690 [Internet]. Washington (DC): U.S. Department of Transportation; September 2005. Available from: https://www.nhtsa.gov/sites/nhtsa.dot.gov/files/medical20cond2080920690-8-04_medical20cond2080920690-8-04.pdf [cited 2019 Jan 7].
[15] Diller E, Cook L, Leonard D, Dean JM, Reading J, Vernon D. Evaluating drivers licensed with medical conditions in Utah, 1992–1996. U.S. DOT/NHTSA Publication No. DOT HS 809 023 [Internet]. Washington (DC): National Highway Traffic Safety Administration; June 1999. Available from: https://one.nhtsa.gov/people/injury/research/utahdrivers/utahmedconditions.html [cited 2019 Jan 7].
[16] Vaa T. Impairment, diseases, age and their relative risks of accident involvement: results from meta-analysis. No. 690/2003 [Internet]. Oslo, Norway: Institute of Transport Economics; December 2003. Available from: https://www.toi.no/getfile.php?mmfileid=5780 [cited 2019 Jan 7].
[17] Hanna R. The contribution of medical conditions to passenger vehicle crashes. U.S DOT/NHTSA Publication No. DOT HS 811 219 [Internet]. Washington (DC): U.S. Department of Transportation; November 2009. Available from: http://www-nrd.nhtsa.dot.gov/Pubs/811219.pdf [cited 2019 Jan 7].
[18] Vingilis E, Wilk P. Medical conditions, medication use, and their relationship with subsequent motor vehicle injuries: examination of the Canadian Population Health Survey. Traffic Inj Prev [Internet] February 2012;13(3):327–36. https://doi.org/10.1080/03081060.2012.635417 [cited 2019 Jan 7].
[19] Rilea SL. Crash risks of drivers with physical and mental (P&M) conditions and changes in crash rates over time. RSS-17-252 [Internet]. Sacramento (CA): California Department of Motor Vehicles; February 2017. Available from: https://trid.trb.org/view/1463963 [cited 2019 Jan 7].
[20] Waller JA. Chronic medical conditions and traffic safety: review of the California experience. N Engl J Med [Internet] December 1965;273(26):1413–20. https://doi.org/10.1056/NEJM196512232732605 [cited 2019 Jan 7].
[21] Insurance Institute of Highway Safety. Highway Loss Data Institute. Older drivers. [Internet]. Arlington (VA): Insurance Institute of Highway Safety; Highway Loss Data Institute; December 2018. Available from: https://www.iihs.org/iihs/topics/t/older-drivers/fatalityfacts/older-people/2017 [cited 2019 Apr 29].
[22] Brown LB, Ott BR. Driving and dementia: a review of the literature. J Geriatr Psychiatry Neurol [Internet] December 2004;17(4):232–40. https://doi.org/10.1177/0891988704269825 [cited 2019 Jan7].
[23] Dawson JD, Anderson SD, Uc EY, Dastrup E, Rizzo M. Predictors of driving safety in early Alzheimer disease. Neurology [Internet] February 2009;72(6):521–7. https://doi.org/10.1212/01.wnl.0000341931.35870.49 [cited 2019 Jan 7].
[24] Dubinsky RM, Stein AC, Lyons K. Practice parameter: risk of driving and Alzheimer's disease (an evidence-based review): report of the quality standards subcommittee of the American Academy of Neurology. Neurology [Internet] June 2000;54(12):2205–11. https://doi.org/10.1212/WNL.54.12.2205 [cited 2019 Jan 7].
[25] Li G, Braver ER, Chen LH. Fragility versus excessive crash involvement as determinants of high death rates per vehicle-mile of travel among older drivers. Accid Anal Prev [Internet] March 2003;35(2):227–35. https://doi.org/10.1016/S0001-4575(01)00107-5 [cited 2019 Jan 7].
[26] Kim S, Ulfarsson GF. Transportation in an aging society: linkage between transportation and quality of life [cited 2019 Jan 7] Transport Res Rec [Internet] January 2013;2357(1):109–15. Available from: http://pascal-francis.inist.fr/vibad/index.php?action=search&terms=27928289.

[27] Ott BR, Festa EK, Amick MM, Grace J, Davis JD, Heindel WC. Computerized maze navigation and on-road performance by drivers with dementia. J Geriatr Psychiatry Neurol [Internet] March 2008;21(1):18–25. https://doi.org/10.1177/0891988707311031 [cited 2019 Jan 7].
[28] Reger MA, Welsh RK, Watson GS, Cholerton B, Baker LD, Craft S. The relationship between neuropsychological functioning and driving ability in dementia: a meta-analysis. Neuropsychology [Internet] January 2004;18(1):85–93. https://doi.org/10.1037/0894-4105.18.1.85 [cited 2019 Jan 7].
[29] Papageorgiou SG, Beratis IN, Kontaxopoulou D, Fragkiadaki S, Pavlou D, Yannis G. Does the diagnosis of Alzheimer's disease imply immediate revocation of a driving license? IJCNMH [Internet] March 2016;3(Suppl. 1):S02. https://doi.org/10.21035/ijcnmh.2016.3(Suppl.1).S02 [cited 2019 Jan 7].
[30] Carr DB, Ott BR. The older adult driver with cognitive impairment: "It's a very frustrating life". JAMA [Internet] April 2010;303(16):1632–41. https://doi.org/10.1001/jama.2010.481 [cited 2019 Jan 7].
[31] Duchek JM, Carr DB, Hunt L, Roe CM, Xiong C, Shah K, Morris JC. Longitudinal driving performance in early-stage dementia of the Alzheimer type. J Am Geriatr Soc [Internet] October 2003;51(10):1342–7. https://doi.org/10.1046/j.1532-5415.2003.51481.x [cited 2019 Jan 7].
[32] Lucas-Blaustein MJ, Filipp L, Dungan C, Tune L. Driving in patients with dementia. J Am Geriatr Soc December 1988;36(12):1087–91.
[33] Hwang Y, Hong GRS. Predictors of driving cessation in community-dwelling older adults: a 3-year longitudinal study. Transp Res Part F Traffoc Psychol Behav [Internet] January 2018;52:202–9. https://doi.org/10.1016/j.trf.2017.11.017 [cited 2019 Apr 11].
[34] Carter K, Monaghan S, O'Brien J, Teodorczuk A, Mosimann U, Taylor JP. Driving and dementia: a clinical decision pathway. Int J Geriatr Psychiatry [Internet] February 2015;30(2):210–6. https://doi.org/10.1002/gps.4132 [cited 2019 Jan 7].
[35] Canadian Council of Motor Transport Administrators. Determining driver fitness in Canada. Part 1: a model for the administration of driver fitness programs. Part 2: CCMTA medical standards for drivers [Internet]. Ottawa, ON: CCMTA; 2017. Available from: https://ccmta.ca/images/pdf-documents-english/CCMTA-Medical-Standards-2017-English.pdf [cited 2019 Jan 8].
[36] Austroads & National Transport Commission, Australia. Assessing fitness to drive for commercial and private vehicle drivers. 5.1 ed. [Internet]. Sydney, AU: Austroads; January 2017. Available from: https://austroads.com.au/__data/assets/pdf_file/0022/104197/AP-G56-17_Assessing_fitness_to_drive_2016_amended_Aug2017.pdf [cited 2019 Jan 8].
[37] New Zealand Transport Agency. Medical aspects of fitness to drive: a guide for health practitioners [Internet]. Wellington, NZ: New Zealand Transport Agency; June 2014. Available from: https://www.nzta.govt.nz/assets/resources/medical-aspects/Medical-aspects-of-fitness-to-drive-A-guide-or-health-practitioners.pdf [cited 2019 Jan 8].
[38] European Commission, Mobility and Transport, 2017 Mobility and Transport Road Safety Fitness to Drive (see https://ec.europa.eu/transport/road_safety/topics/behaviour/fitness_to_drive_en).
[39] Canadian Medical Association. CMA driver's guide: determining medical fitness to operate motor vehicles [Internet]. 9th ed. Ottawa (ON): Joule, Inc; 2017. Available from: https://joulecma.ca/evidence/CMA-drivers-guide [cited 2019 Jan 8].
[40] American Geriatrics Society. Clinician's guide to assessing and counseling older drivers [Internet]. 3rd ed. Washington, DC: National Highway Traffic Safety Administration; January 2016. Available from: https://www.nhtsa.gov/sites/nhtsa.dot.gov/files/812228_cliniciansguidetoolderdrivers.pdf [cited 2019 Jan 8].

[41] Driver & Vehicle Licensing Agency. Assessing fitness to drive: a guide for medical professionals. [Internet]. Swansea (UK): The Department for Transport; August 2018. Available from: https://assets.publishing.service.gov.uk/government/uploads/system/uploads/attachment_data/file/736938/assessing-fitness-to-drive-a-guide-for-medical-professionals.pdf [cited 2018 Jan 8].
[42] Brubacher J, Renschler C, Gomez AM, Huang B, Lee CW, Erdely S, Chan H, Purssell R. Reporting unfit drivers: knowledge, attitudes, and practice of BC physicians [Internet]. Vancouver (BC): University of British Columbia; 2018. Available from: http://med-fom-emerg.sites.olt.ubc.ca/files/2018/03/Fitness-to-Drive-Survey-final-2018-03-06.pdf [cited 2019 Jan 10].
[43] Alkharboush GA, Al Rashed FA, Saleem AH, Alnajashi IS, Almeneessier AS, Olaish AH, et al. Assessment of patients' medical fitness to drive by primary care physicians: a cross-sectional study. Traffic Inj Prev [Internet] July 2017;18(5):488—9. https://doi.org/10.1080/15389588.2016.1274029 [cited 2019 Jan 10].
[44] Gergerich EM. Reporting policy regarding drivers with dementia. Gerontologist [Internet] April 2016;56(2):345—56. https://doi.org/10.1093/geront/gnv143 [cited 2019 Jan 10].
[45] Meuser TM, Carr DB, Ulfarsson GF. Motor-vehicle crash history and licensing outcomes for older drivers reported as medically impaired in Missouri. Accid Anal Prev [Internet] March 2009;41(2):246—52. https://doi.org/10.1016/j.aap.2008.11.003 [cited 2019 Jan 7].
[46] Jang RW, Man-Son-Hing M, Naglie G, Molnar FJ, Hogan DB, Marshall SC, et al. Family physicians' attitudes and practices regarding assessments of medical fitness to drive in older persons. J Gen Intern Med [Internet] April 2007;22(4):531—43. https://doi.org/10.1007/s11606-006-0043-x [cited 2019 Jan 10].
[47] Louie AV, D'Souza DP, Palma DA, Bauman GS, Lock M, Fisher B, Patil N, Rodrigues GB. Fitness to drive in patients with brain tumours: the influence of mandatory reporting legislation on radiation oncologists in Canada. Curr Oncol [Internet] June 2012;19(3):e117—22. https://doi.org/10.3747/co.19.916 [cited 2019 Jan 10].
[48] Elgar NJ, Smith BJ. Mandatory reporting by doctors of medically unsafe drivers is unpopular and poorly adhered to: a survey of sleep physicians and electro-physicians. Intern Med J [Internet] March 2018;48(3):293—300. https://doi.org/10.1111/imj.13620 [cited 2019 Jan 10].
[49] Carr DB, Schwartzberg JG, Manning L, Sempek K. Physician's guide to assessing and counseling older drivers [Internet]. 2nd ed. Washington (DC): National Highway Traffic Safety Administration; 2010. Available at: http://www.ama-assn.org/ama1/pub/upload/mm/433/older-drivers-guide.pdf [cited 2019 Apr 29].
[50] Redelmeier DA, Yarnell CJ, Thiruchelvam D, Tibshirani RJ. Physicians' warnings for unfit drivers and the risk of trauma from road crashes. N Engl J Med [Internet] September 2012;367(13):1228—36. https://doi.org/10.1056/NEJMsa1114310 [cited 2019 Jan 10].
[51] Sinnott C, Foley T, Forsyth J, McLoughlin K, Horgan L, Bradley CP. Consultations on driving in people with cognitive impairment in primary care: a scoping review of the evidence. PLoS One [Internet] October 2018;13(10):e0205580. https://doi.org/10.1371/journal.pone.0205580 [cited 2019 Apr 11].
[52] Kahvedžić A, Mcfadden R, Cummins G, Carr D, O'Neill D. Impact of new guidelines and educational programs on awareness of medical fitness to drive among general practitioners in Ireland. Traffic Inj Prev [Internet] August 2014;16(6):593—8. https://doi.org/10.1080/15389588.2014.979408 [cited 2019 Jan 11].

[53] Weir E. Using the Upshur principles to discuss medical fitness to drive [cited 2019 Jan 10] Can Fam Physician [Internet] April 2017;vol. 63(4):269–72. Available from: http://www.cfp.ca/content/63/4/269.

[54] Berger JT, Rosner F, Kark P, Bennett AJ. Reporting by physicians of impaired drivers and potentially impaired drivers. J Gen Intern Med [Internet September 2000;15(9):667–72. https://doi.org/10.1046/j.1525-1497.2000.04309.x [cited 2019 Jan 9].

[55] Government of Alberta. Reporting concerns about driver fitness [Internet]. Edmonton (AB): Driver Fitness and Monitoring, Alberta Transportation; 2019. Available from: https://www.transportation.alberta.ca/2561.htm [cited 2019 Jan 9].

[56] Meuser TM, Carr DB, Unger EA, Ulfarsson GF. Family reports of medically impaired drivers in Missouri: cognitive concerns and licensing outcomes. Accid Anal Prev [Internet] January 2015;74:17–23. https://doi.org/10.1016/j.aap.2014.10.002 [cited 2019 Jan 7].

[57] Dugan E, Barton KN, Coyle C, Man Lee CUS. Policies to enhance older driver safety: a systematic review of the literature. J Aging Soc Policy [Internet] September 2013;25(4):335–52. https://doi.org/10.1080/08959420.2013.816163 [cited 2019 May 7].

[58] Moore A, Frank C, Chambers LW. Role of the family physician in dementia care [cited 2019 Jan 6] Can Fam Physician October 2018;64(10):717–9. Available from: http://www.cfp.ca/content/64/10/717.

[59] Troisième conférence canadienne de consensus sur le diagnostic et le traitement de la démence [Internet]. Montréal (QC): Conférence Canadienne de Consensus sur le Diagnostic et le Traitement de la Démence; September 2007. Available from: http://www.cccdtd.ca/pdfs/recommandations_approuvees_cccdtd_2007.pdf [cited 2019 Jan 6].

[60] Organizing Committee. Canadian consensus conference on the assessment of dementia. Assessing dementia: the Canadian consensus. CMAJ April 1991;144(7):851–3. Available from: https://www.ncbi.nlm.nih.gov/pmc/articles/PMC1335280/ [cited 2019 Jan 6].

[61] Moore A, Patterson C, Lee L, Vedel I, Bergman H. Fourth Canadian consensus on the diagnosis and treatment of dementia: recommendations for family physicians [cited 2019 Jan 6] Can Am Physician [Internet] May 2014;60(5):433–8. Available from: https://www.ncbi.nlm.nih.gov/pmc/articles/PMC4020644/.

[62] Canadian Institute for Health Information. Family doctor preparedness [Internet]. Ottawa, ON: CIHI; June 2018. Available from: https://www.cihi.ca/en/dementia-in-canada/spotlight-on-dementia-issues/family-doctor-preparedness [cited 2019 Jan 8].

[63] Carr DB. The older adult driver [cited 2019 Jan 11] Am Fam Physician [Internet] January 2000;61(1):141–6. Available from: https://www.aafp.org/afp/2000/0101/p141.html.

[64] Hum S, Cohen C, Persaud M, Lee J, Drummond N, Dalziel W, Pimlott N. Role expectations in dementia care among family physicians and specialists. Can Geriatr J [Internet] September 2014;17(3):95–102. https://doi.org/10.5770/cgj.17.110 [cited 2019 Jan 11].

[65] Dal Bello-Haas VPM, Cammer A, Morgan D, Stewart N, Kosteniuk J. Rural and remote dementia care challenges and needs: perspectives of formal and informal care providers residing in Saskatchewan, Canada. Rural Remote Health July 2014;14(3):1–13.

[66] Turner S, Iliffe S, Downs M, Wilcock J, Bryans M, Levin E, Keady J, O'Carroll R. General practitioner's knowledge, confidence and attitudes in the diagnosis and management of dementia. Age Ageing [Internet] July 2004;33(5):461–7. https://doi.org/10.1093/ageing/afh140 [cited 2019 Jan 11].

[67] Moorehouse P, Hamilton L, Fisher T, Rockwood K. Barriers to assessing fitness to drive in dementia in Nova Scotia: informing strategies for knowledge translation. Can Geriatr J [Internet] September 2011;14(3):61–5. https://doi.org/10.5770/cgj.v14i3.7 [cited 2019 Apr 11].

[68] Folstein MF, Folstein SE, McHugh PR. 'Mini-mental state': a practical method for grading the cognitive state of patients for the clinician. J Psychiatr Res [Internet] November 1975;12(3):189–98. https://doi.org/10.1016/0022-3956(75)90026-6 [cited 2019 Jan 8].

[69] Nasreddine ZS, Phillips NA, Bédirian V, Charbonneau S, Whitehead V, Collin I. Chertkow. The Montreal cognitive assessment, MoCA: a brief screening tool for mild cognitive impairment. J Am Geriatr Soc [Internet] March 2005;53(4):695–9. https://doi.org/10.1111/j.1532-5415.2005.53221.x [cited 2019 Jan 8].

[70] Reitan RM. Validity of the trail making test as an indicator of organic brain damage. Percept Mot Skills [Internet] December 1958;8:271–6. https://doi.org/10.2466/pms.1958.8.3.271 [cite 2019 Jan 8].

[71] Shulman KI. Clock-drawing: is it the ideal cognitive screening test? Int J Geriatr Psychiatry [Internet] June 2000;15(6):548–61. https://doi.org/10.1002/1099-1166(200006)15:6<548::AID-GPS242>3.0.CO;2-U [cited 2019 Jan 8].

[72] Bennett JM, Chekaluk E, Batchelor J. Cognitive tests and determining fitness to drive in dementia: a systematic review [cited 2019 Jan 10] J Am Geriatr Soc [Internet] September 2016;vol. 64(9). https://doi.org/10.1111/jgs.14180. 1904–17.

[73] Dobbs BM, Schopflocher D. The introduction of a new screening tool for the identification of cognitively impaired medically at-risk drivers. J Prim Care Commun Health [Internet] July 2010;1(2):119–27. https://doi.org/10.1177/2150131910369156 [cited 2019 Jan 8].

[74] Roe CM, Babulal GM, Head DM, Stout SH, Vernon EK, Ghoshal N, et al. Preclinical Alzheimer's disease and longitudinal driving decline. Alzehimers Dement: TRCI [Internet] January 2017;3(1):74–82. https://doi.org/10.1016/j.trci.2016.11.006 [cited 2019 Jan 10].

[75] Stout SH, Babulal GM, Ma C, Head DM, Grant EA, Holtzman DM, et al. Driving cessation over a 24-year period: dementia severity and cerebrospinal fluid biomarkers. Alzheimers Dement [Internet] May 2018;14(5):610–6. https://doi.org/10.1016/j.jalz.2017.11.011 [cited 2019 Apr 11].

[76] Bixby K, David JD, Ott BR. Comparing caregiver and clinician predictions of fitness to drive in people with Alzheimer's disease. Am J Occup Ther [Internet] May–June 2016. https://doi.org/10.5014/ajot.2015.013631 [cited 2019 Apr 15].

[77] Cotrell V, Wild K. Longitudinal study of self-imposed driving restrictions and deficit awareness in patients with Alzheimer's disease. Alzheimer Dis Assoc Disord [Internet] July/September 1999;13(5):152–6. https://doi.org/10.1097/00002093-199907000-00007 [cited 2019 Apr 14].

[78] Wild K, Cotrell V. Identifying driving impairment in Alzheimer disease: a comparison of self and observer reports versus driving evaluation. Alzheimer Dis Assoc Disord [Internet] January 2003;17(1):27–34. https://doi.org/10.1097/00002093-200301000-00004 [cited 2019 Apr 14].

[79] Brown L, Ott B, Papandonatos G, Sui Y, Ready R, Morris J. Predictors of on-road driving performance in patients with early Alzheimer's disease. J Am Geriatr Soc [Internet] January 2005;52(1):94–8. https://doi.org/10.1111/j.1532-5415.2005.53017 [cited 2019 Apr 14].

[80] Mauri M, Sinforiani E, Cuzzoni MG, Bono G, Zucchella C. Driving habits in patients with dementia: a report from Alzheimer's disease assessment units in northern Italy. Funct Neurol [Internet] April–June 2014;29(2):107–12. https://doi.org/10.11138/FNeur/2014.29.2.107 [cited 2019 Apr 11].

[81] Odenheimer GL, Beaudet M, Jette AM, Albert MS, Grande L, Minaker KL. Performance-based driving evaluation of the elderly driver – safety, reliability, and validity. J Gerontol [Internet July 1994;49(1):M153–9. https://doi.org/10.1093/geronj/49.4.M153 [cited 2019 Jan 9].

[82] Pinner G, Wilson S. Driving and dementia: a clinician's guide. Adv in Psychiatr Treat [Internet] March 2013;19(2):89–96. https://doi.org/10.1192/apt.bp.111.009555 [cited 2019 Apr 11].

[83] Piersma D, Fuermaier ABM, de Waard D, De Deyn PP, Davidse RJ, de Groot J, et al. The MMSE should not be the sole indicator of fitness to drive in mild Alzheimer's dementia. Acta Neurol Belgica [Internet] December 2018;118(4):637–42. https://doi.org/10.1007/s13760-018-1036-3 [cited 2019 Jan 9].

[84] Allan CL, Behrman S, Baruch N, Ebmeier KP. Driving and dementia: a clinical update for mental health professionals. Evid Based Ment Health [Internet] November 2016;19(4):110–3. https://doi.org/10.1136/eb-2016-102485 [cited 2019 Apr 11].

[85] Alzheimer's Association. New analysis shows more than 28 million Baby Boomers will develop Alzheimer's disease; will consume nearly 25% of medicare spending [Internet]. Washington (DC): Alzheimer's Association; July 20, 2015 [cited 2019 Apr 155]. Available from: https://www.alz.org/aaic/_downloads/mon-930am-baby-boomers.pdf.

[86] Lee L, Molnar F. Driving and dementia [internet]. Can Fam Physician January 2017, 2017;63(1):27–31. PMC5257216/.

[87] Swaffeer K. Dementia and the impact of not driving [Internet]. Wollongong (AUS): Australian Journal of DementiaCare; September 4, 2013. Available from: https://journalofdementiacare.com/dementia-and-driving/ [cited 2019 Apr 15].

[88] Yaffe MJ, Orzeck P, Barylak L. Family physicians' perspectives on care of dementia patients and family caregivers. Can Fam Physician [Internet] July 2008;54(7):1008–15. Available from: http://www.cfp.ca/content/54/7/1008 [cited 2019 Jan 11].

[89] Alzheimer's Research UK. Dementia in the family: the impact on carers. [Internet]. Cambridge: Alzheimer's Research UK; December 2015. Available from: https://www.alzheimersresearchuk.org/about-us/our-influence/reports/carers-report/ [cited 2019 Apr 11].

[90] Byszewski A, Molnar FJ, Aminzadeh F. The impact of disclosure of unfitness to drive in persons with newly diagnosed dementia: patient and caregiver perspectives. Clin Gerontol [Internet] April 2010;33(2):152–63. https://doi.org/10.1080/07317110903552198 [cited 2019 Apr 11].

[91] Ng KY, Garnham J, Syed UM, et al. Knowledge of driving vehicle licensing agency guidelines among NHS doctors: a multicentre observational study. JRSM Open 2015;6(10). https://doi.org/10.1177/2054270415601586.

[92] Omer S, Dolan C, Dimitrov BD, Langan C, McCarthy G. General practitioners' opinions and attitudes towards medical assessment of fitness to drive of older adults in Ireland. Austalas J Ageing [Intenet] May 2013;33(3):E33–8. https://doi.org/10.1111/ajag.12045.

[93] Hakamies-Blomqvist L, Henriksson P, Falkmer T, Lundberg C, Braekhus A. Attitudes of primary care physicians toward older drivers: a Finnish-Swedish comparison. J Appl Gerontol [Internet] March 2002;21(1):58–69. https://doi.org/10.1177/073346480202100105 [cited 2019 Jan 11].

[94] Morris JC, Perkinson MA, Berg-Weger ML, Carr DB, Meuser TM, Palmer JL, et al. Driving and dementia of the Alzheimer type: beliefs and cessation strategies among stakeholders. Gerontologist [Internet] October 2005;45(5):676–85. https://doi.org/10.1093/geront/45.5.676 [cited 2019 Jan 11].
[95] Odenheimer GL. Driver safety in older adults. The physician's role in assessing driving skills of older patients. Geriatrics October 2006;61(10):14–21.
[96] Hamdy RC, Kinser A, Kendall-Wilson T, Depelteau A, Whalen K, Culp J. Driving and patients with dementia. Gerontol Geriatr Med [Internet] January 2018;4. https://doi.org/10.1177/2333721418777085 [cited 2019 Jan 11].
[97] Eisenhandler SA. The asphalt identikit: old age and the driver's license. Int J Aging Hum Dev [Internet] January 1990;30(1):1–14. https://doi.org/10.2190/0MF5-HQ1L-7EBY-XNXV [cited 2019 Apr 11].
[98] Breen DA, Breen DP, Moore JW, Breen PA, O'Neill D. Driving and dementia. BMJ [Internet] June 2007;334(7608):1365–9. https://doi.org/10.1136/bmj.39233.585208.55 [cited 2019 Jan 7].
[99] Rudman DL, Friedland J, Chipman M, Sciortino P, Brayne C, Chipman ML, et al. Holding on and letting go: the perspectives of pre-seniors and seniors on driving self-regulation in later life. Can J Aging [Internet] March 2006;25(1):65–76. https://doi.org/10.1353/cja.2006.0031 [cited 2019 Apr 11].
[100] Dong Y, Gan DZ, Tay SZ, Koay WI, Collinson SL, Hilal S, et al. Patterns of neuropsychological impairment in Alzheimer's disease and mixed dementia. J Neurol Sci [Internet] October 2013;333(1–2):5–8. https://doi.org/10.1016/j.jns.2013.05.011 [cited 2019 Apr 11].
[101] Levy JA, Chelune GJ. Cognitive-behavioral profiles of neurodegenerative dementias: beyond Alzheimer's disease. J Geriatr Psychiatry Neurol [Internet] December 2007;20(4):227–38. https://doi.org/10.1177/0891988707308806 [cited 2019 Apr 11].
[102] Gilley DW, Wilson RS, Bennett DA, Stebbins GT, Bernard BA, Whalen ME, et al. Cessation of driving and unsafe motor vehicle operation by dementia patients. Arch Intern Med [Internet] May 1991;151(5):941–6. https://doi.org/10.1001/archinte.1991.00400050087017 [cited 2019 Apr 11].
[103] Seiler S, Schmidt H, Lechner A, Benke T, Sanin G, Ransmayr G, et al. Driving cessation and dementia: results of the prospective registry on dementia in Austria (PRODEM). PLoS One December 2012;7(12):e52710. https://doi.org/10.1371/journal.pone.0052710 [cited 2019 Apr 11].
[104] Foley DJ, Masaki KH, Ross GW, White LR. Driving cessation in older men with incident dementia. J Am Geriatr Soc [Internet] August 2000;48(8):928–30. https://doi.org/10.1111/j.1532-5415.2000.tb06889.x [cited 2019 Apr 11].
[105] Talbot A, Bruce I, Cunningham CJ, Coen RF, Lawlor BA, Coakley D, Walsh JB, O'Neill D. Driving cessation in patients attending a memory clinic. Age Ageing [Internet] July 2005;34(4):363–8. https://doi.org/10.1093/ageing/afi090 [cited 2019 Apr 11].
[106] Allan CL, Behrman S, Baruch N, et al. Driving and dementia: a clinical update for mental health professionals. Evid Based Ment Health 2016;19:110–3. doi.org/10.1136/eb-2016-102485.
[107] Amieva H, Jacqmin-Gadda H, Orgogozo JM, et al. The 9 year cognitive decline before dementia of the Alzheimer type: a prospective population-based study. Brain [Internet] May 2005;128(5):1093–101. https://doi.org/10.1093/brain/awh451 [cited 2019 Apr 11].
[108] Kawas CH, Corrada MM, Brookmeyer R, et al. Visual memory predicts Alzheimer's disease more than a decade before diagnosis. Neurology [Internet] April 2003;60(7):1089–93. https://doi.org/10.1212/01.WNL.0000055813.36504.BF [cited 2019 Apr 11].

[109] Rajan KB, Wilson RS, Weuve J, Barnes LL, Evans DA. Cognitive impairment 18 years before clinical diagnosis of Alzheimer disease dementia. Neurology [Internet] September 2015;85(10):898–904. https://doi.org/10.1212/WNL.0000000000001774 [cited 2019 Apr 11].
[110] Marie Dit Asse L, Fabrigoule C, Helmer C, Laumon B, Berr C, Rouaud O, et al. Gender effect on driving cessation in pre-dementia and dementia phases: results of the 3C population-based study. Int J Geriatr Psychiatry[Internet] October 2017;32(10):1049–58. https://doi.org/10.1002/gps.4565 [cited 2019 Jan 11].
[111] Velayudhan L, Baillon S, Urbaskova G, McCulloch L, Tromans S, Storey M, Lindesay J, Bhattacharyya S. Driving cessation in patients attending a young-onset dementia clinic: a retrospective cohort study. Dement Geriatr Cogn Dis Extra [Internet] April 2018;8(1):190–8. https://doi.org/10.1159/000488237 [cited 2019 Apr 11].
[112] Connors MH, Ames D, Woodward M, Brodaty H. Predictors of driving cessation in dementia: baseline characteristics and trajectories of disease progression. Alzheimer Dis Assoc Disord [Internet] 2018;32(1):57–61. https://doi.org/10.1097/WAD.0000000000000212 [cited 2019 Jan 7].
[113] Herrmann N, Rapoport MJ, Sambrook R, Hébert R, McCracken P, Robillard A. Predictors of driving cessation in mild-to-moderate dementia. CMAJ [Internet] September 2006;175(6):591–5. https://doi.org/10.1503/cmaj.051707 [cited 2019 Apr 11].
[114] Carr DB, Shead V, Storandt M. Driving cessation in older adults with dementia of the Alzheimer's type. Gerontologist [Internet] December 2005;45(6):824–7. https://doi.org/10.1093/geront/45.6.824 [cited 2019 Apr 11].
[115] Wilson RS, Sytsma J, Barnes LL, Boyle PA. Anosognosia in dementia. Curr Neurol Neurosci Rep [Internet] September 2016;16(9):77. https://doi.org/10.1007/s11910-016-0684-z [cited 2019 Apr 11].
[116] Starkstein SE, Chemerinski E, Sabe L, et al. Prospective longitudinal study of depression and anosognosia in Alzheimer's disease. Br J Psychiatry [Internet] July 1997;171(7):47–52. https://doi.org/10.1136/jnnp.2005.069575 [cited 2019 Apr 11].
[117] Lin F, Wharton W, Dowling NM, et al. Awareness of memory abilities in community-dwelling older adults with suspected dementia and mild cognitive impairment. Dement Geriatr Cogn Disord [Internet] August 2010;30(1):83–92. https://doi.org/10.1159/000318752 [cited 2019 Apr 11].
[118] Fragkiadaki S, Beratis IN, Kontaxopoulou D, Papadimitriou E, Andronas N, Yannis G, Papageorgiou SG. Self-awareness of cognitive and driving ability in patients with Mild Cognitive Impairment, Alzheimer's disease and healthy elderly. Front Hum Neurosci Conference Abstract: SAN2016 Meeting [Internet] October 2016. https://doi.org/10.3389/conf.fnhum.2016.220.00114 [cited 2019 Apr 11].
[119] Okonkwo OC, Griffith HR, Vance DE, Marson DC, Ball KK, Wadley VG. Awareness of functional difficulties in mild cognitive impairment: a multidomain assessment approach. J Am Geriatr Soc [Internet] June 2009;57(6):978–84. https://doi.org/10.1111/j.1532-5415.2009.02261.x [cited 2019 Apr 11].
[120] Brown LB, Ott BR, Papandonatos GD, Sui Y, Ready RE, Morris JC. Prediction of on road driving performance in patients with early Alzheimer's disease. J Am Geriatr Soc [Internet] January 2005;53(1):94–8. https://doi.org/10.1111/j.1532-5415.2005.53017.x [cited 2019 Apr 11].
[121] Wong I, Smith S, Sullivan K. The relationship between cognitive ability, insight and self-regulatory behaviors: findings from the older driver population. Accid Anal Prev [Internet] November 2012;49:316–21. https://doi.org/10.1016/j.aap.2012.05.031 [cited 2019 Apr 11].

[122] Mizuno Y, Arai A, Arai Y. Determination of driving cessation for older adults with dementia in Japan. Int J Geriatr Psychiatry [Internet] September 2008;23(9):987–9. https://doi.org/10.1002/gps.1999 [cited 2019 Apr 11].

[123] Liddle J, Haynes M, Pachana NA, Mitchell G, McKenna K, Gustafsson L. Effect of a group intervention to promote older adults' adjustment to driving cessation on community mobility: a randomized controlled trial. Gerontologist [Internet] June 2014;54(3):409–22. https://doi.org/10.1093/geront/gnt019 [cited 2019 Apr 11].

[124] Croston J, Meuser TM, Berg-Weger M, Grant EA, Carr DB. Driving retirement in older adults with dementia. Top Geriatr Rehabil [Internet] April 2009;25(2): 154–62. https://doi.org/10.1097/TGR.0b013e3181a103fd [cited 2019 Apr 11].

[125] Taylor BD, Tripodes S. The effects of driving cessation on the elderly with dementia and their caregivers. Accid Anal Prev [Internet July 2001;33(4):519–28. https://doi.org/10.1016/S0001-4575(00)00065-8 [cited 2019 Apr 11].

[126] Liddle J, Bennett S, Allen S, Lie DC, Standen B, Pachana NA. The stages of driving cessation for people with dementia: needs and challenges. Int Psychogeriatr [Internet] 2013 Dec;25(12):2033–46. https://doi.org/10.1017/S1041610213001464 [cited 2019 Apr 11].

[127] Friedland RP, Koss E, Kumar A, Gaine S, Metzler D, Haxby JV, et al. Motor vehicle crashes in dementia of the Alzheimer type. Ann Neurol [Internet] December 1988;24(6):782–6. https://doi.org/10.1002/ana.410240613 [cited 2019 Apr 11].

[128] Adler G, Kuskowski M. Driving cessation in older men with dementia. Alzheimer Dis Assoc Disord [Internet] April 2003;17(2):68–71. https://doi.org/10.1097/00002093-200304000-00003 [cited 2019 Apr 14].

[129] Chihuri S, Mielenz TJ, Li G, Dimaggio CJ, Betz ME, Diguiseppi C, et al. Driving cessation and health outcomes in older adults. J Am Geriatr Soc [Internet] February 2016;64(2):332–41. https://doi.org/10.1111/jgs.13931 [cited 2019 Apr 15].

[130] Betz ME, Lowenstein SR, Schwartz R. Older adult opinions of "advance driving directives". J Prim Care Commun Health [Internet] January 2013;4(1):14–27. https://doi.org/10.1177/2150131912447082 [cited 2019 Jan 11].

[131] The Hartford Center for Mature Market Excellence. At the crossroads: family conversations about Alzheimer's disease, dementia & driving [Internet]. Hartford, CT: The Hartford Financial Services Group, Inc.; August 2015. Available from: https://www.thehartford.com/resources/mature-market-excellence/publications-on-aging [cited 2019 Jan 8].

[132] Dobbs B, Pidborochynski T, Linsdell M. The senior driver myths & facts: information for health care professionals, patients, and families on assessment and referral issues. 3rd ed. Edmonton (AB): The DR Group; 2015.

[133] Seguin D. The experience of driving cessation in dementia: examples from Ontario and Alberta [master's thesis on the internet]. Ottawa: University of Ottawa; 2014. https://doi.org/10.20381/ruor-6584 [cited 2019 Apr 14]. Available from: https://ruor.uottawa.ca/handle/10393/31550.

[134] Musselwhite C, Haddad H. Mobility, accessibility and quality of later life. Qual Ageing Older Adults [Internet] March 2010;11(1):25–37. https://doi.org/10.5042/qiaoa.2010.0153 [cited 2019 Apr 14].

[135] Dobbs BM. Transitioning from driving to driving cessation: results from a two year study evaluating driving cessation support groups for individuals with dementia and their family members. Poster presented at: the 5th Annual Family Medicine Research Day; June 8, 2007. Edmonton, (AB).

[136] Chacko EE, Wright WM, Worrall RC, Adamson C, Cheung G. Reactions to driving cessation: a qualitative study of people with dementia and their families. Australas Psychiatry [Internet] June 2015;23(5):496–9. https://doi.org/10.1177/1039856215591326 [cited 2019 Apr 29].

[137] Alzheimer Society of Canada. Conversations about driving and dementia [internet]. Toronto (ON): Alzheimer Society of Canada; November 2017. Available from: https://alzheimer.ca/sites/default/files/files/national/brochures-conversations/conversations-about-dementia-and-driving.pdf [cited 2019 Jan 11].

[138] Fry D, Fox B, Donnelly C. Traveling a new road: a driving cessation group in primary care. Occup Ther Now September 2013;15(5):25–6 [cited 2019 Apr 14].

[139] Pinquart M, Sörensen S. Helping caregivers of persons with dementia: which interventions work and how large are their effects? Inter Psychogeriatr [Internet] December 2006;18(4):577–95. https://doi.org/10.1017/S1041610206003462 [cited 2019 Apr 15].

[140] Chien LY, Chu H, Guo JL, Liao YM, Chang LI, Chen CH, Chou KR. Caregiver support groups in patients with dementia: a meta-analysis. Int J Geriatr Psychiatry [Internet] October 2011;26(10):1089–98. https://doi.org/10.1002/gps.2660 [cited 2019 Apr 14].

[141] Stern RA, D'Ambrosio LA, Mohyde M, Carruth A, Tracton Bishop B, Hunter JC, et al. At the crossroads: development and evaluation of a dementia caregiver group intervention to assist in driving cessation. Gerontol Geriatr Educ [Internet] December 2008;29(4):363-382. https://doi.org/10.1080/02701960802497936 [cited 2019 Apr 14].

[142] Lazarus RS, Folkman S. Stress, appraisal, and coping. New York: Springer Publishing Company; 1984.

[143] Dobbs BM, Harper LA, Wood A. Transitioning from driving to driving cessation: the role of specialized driving cessation support groups for individuals with dementia. Top Geriatr Rehabil [Internet] January 2009;25(1):73–86. https://doi.org/10.1097/01.TGR.0000346058.32801.95 [cited 2019 Apr 15].

[144] Centers for Disease Control and Prevention. The state of aging and health in America [Internet]. Atlanta (GA): Centers for Disease Control and Prevention, US Department of Health and Human Services; 2013. Available from: https://www.cdc.gov/aging/pdf/State-Aging-Health-in-America-2013.pdf [cited 2019 Apr 14].

[145] Florida Department of Elder Affairs. Assessing the needs of elder Floridians. [Internet]. Tallahassee (FL): Florida Department of Elder Affairs; 2016. Available from: http://elderaffairs.state.fl.us/doea/pubs/pubs/2016_Assessing_the_Needs_of_Elder_Floridians.pdf [cited 2019 Apr 14].

[146] Vanderbug M, Silverstein NM. Community mobility and dementia. A review of the literature. DOT HS 810 684. [Internet]. Washington (DC): Alzheimer's Association Public Policy Division and The National Highway Traffic Safety Administration; November 2006. Available from: https://one.nhtsa.gov/people/injury/olddrive/CommMobilityDementia/CommMobileandDementia.pdf [cited 2019 May 7].

[147] Brown JR, Duncan M, Horner MW, Bond M, Wood J. Provider perspectives on six strategies to overcome the barriers to older adult use of alternative transportation services: evidence from seven communities. Case Stud Transp Policy [Internet] June 2018;6(2):237–45. https://doi.org/10.1016/j.cstp.2018.04.004 [cited 2019 Apr 14].

[148] Dobbs BM, Bhardwaj P, Pidborochynski T. Alternate transportation for seniors. An examination of service providers in urban and rural Alberta – executive summary. Edmonton, AB: The DR Group; 2010.

[149] Miller J, Ward C, Lee C, D'Ambrosio L, Coughlin J. Sharing is caring: the potential of the sharing economy to support aging in place. Gerontol Geriatr Educ [Internet] January 2018:1–23. https://doi.org/10.1080/02701960.2018.1428575 [cited 2019 Apr 15].

[150] Pew Research Centre. Tech adoption climbs among older adults. [Internet]. Washington (DC): Pew Research Centre; May 2017. Available from: https://www.pewinternet.org/2017/05/17/technology-use-among-seniors/ [cited 2019 Apr 14].

[151] Duncan M, Charness N, Chapin T, Horner M, Stevens L, Richard A, et al. Enhanced mobility for aging populations using automated vehicles [Internet]. Tallahassee (FL): Florida Department of Transportation; December 2015. Available from: https://ftp.fdot.gov/file/d/FTP/FDOT%20LTS/CO/research/Completed_Proj/Summary_PL/FDOT-BDV30-977-11-rpt.pdf [cited 2019 Apr 14].
[152] Kyriakidis M, Happee R, De Winter JCF. Public opinion on automated driving: results of an international questionnaire among 5,000 respondents. Transp Res F Traffic Psychol Behav [Internet] July 2015;32:127–40. https://doi.org/10.1016/j.trf.2015.04.014 [cited 2019 Apr 14].
[153] Abraham H, Reimer B, Seppelt B, Fitzgerald C, Mehler B, Coughlin JF. Consumer interest in automation: change over one year. In: Proceedings of the Transportation Research Board 97th Annual Meeting; January 7–11, 2018 [Washington (DC)].
[154] Eby DW, Molanr LJ, Stanciu SC. Older adults' attitudes and opinions about automated vehicles: a literature review. Report: ATLAS-2018-26. Ann Arbor (MI): University of Michigan Center for Advancing Transportation Leadership and Safety (ATLAS Center); November 2018. Available from: http://www.atlas-center.org/wp-content/uploads/2018/11/ATLAS-Report-Older-Adults-and-Autonomous-Vehicles.pdf [cited 2019 Apr 14].
[155] Denton M, Ploeg J, Tindale J, Hutchison B, Brazil K, Akhtar-Danesh N, et al. Where would you turn for help? Older adults' awareness of community support services. Can J Aging [Internet] 2008;27(4):359–70. https://doi.org/10.3138/cja.27.4.359 [cited 2019 Jan 11].
[156] Ploeg J, Denton M, Tindale J, Hutchison B, Brazil K, Akhtar-Danesh N, et al. Older adults' awareness of community health and support services for dementia care. Can J Aging [Internet] December 2009;28(4):359–70. https://doi.org/10.1017/S0714980809990195 [cited 2019 Jan 11].
[157] The Hartford Center for Mature Market Excellence. We need to talk: family conversations with older drivers [Internet]. Hartford, CT: The Hartford Financial Services Group, Inc.; May 2018. Available from: https://www.thehartford.com/resources/mature-market-excellence/publications-on-aging [cited 2019 Jan 8].
[158] Alzheimer's Disease International. Dementia friendly communities [internet]. London: Alzheimer's Disease International; 2019. Available from: https://www.alz.co.uk/dementia-friendly-communities [cited 2019 May 7].
[159] Wolfe A. Dementia friendly communities: municipal toolkit [Internet]. Regina (SK): Alzheimer Society Saskatchewan. Available from: http://www.dementiafriendlysaskatchewan.ca/assets/dfc_municipal_toolkit_web.pdf.
[160] Motamedi M, Hussey E, Dobbs B. A guide to mobility and independence: a comprehensive listing of essential services and transportation providers in Edmonton and outlying communities. 9th ed. Edmonton (AB): The DR Group; 2019.
[161] Hussey E, Motamedi M, Dobbs B. A guide to mobility and independence: a comprehensive listing of essential services and transportation providers in Calgary and outlying communities. 3rd ed. Edmonton (AB): The DR Group; 2019.
[162] Medically At-Risk Driver Centre. Provincial listing of organizations providing alternate transportation for seniors in Alberta [Internet]. Edmonton (AB): Medically At-Risk Driver Centre; 2018. Available from: https://mard.med.ualberta.ca/mard/db/ [cited 2019 Jan 10].

CHAPTER 4

Transitions in financial autonomy and risk for financial elder abuse

Stacey Wood[1], Peter A. Lichtenberg[2], Muncuran Purewal[3], Linda Garcia[4,5]
[1]Scripps College, Claremont, CA, United States; [2]Institute of Gerontology and Merrill Palmer Skillman Institute, Wayne State University, Detroit, MI, United States; [3]Department of Family Medicine, Faculty of Medicine, University of Ottawa, Ottawa, Canada; [4]Life Research Institute, University of Ottawa, Ottawa, ON, Canada; [5]Interdisciplinary School of Health Sciences, Faculty of Health Sciences, University of Ottawa, Ottawa, ON, Canada

Case example

Charles is an 80-year-old Caucasian man who has been the caregiver for his wife, Helen, for about 2 years. Charles had a traditional "breadwinner" role, and Helen managed the day-to-day budgeting and finances. Charles reported that they always discussed major financial decisions in their marriage and had used the same broker for years for investments. After years of financial stability, Charles has grown anxious regarding the potential cost of in-home caregivers to assist with Helen. Charles and Helen have two adult sons, David and Thomas, who live about an hour away. They are concerned about their parents' financial well-being and call to check in about once per week. Helen had served as the social bridge for the couple and was an enormous source of emotional support for the whole family. When his sons call, Charles has a tendency to minimize their challenges, indicating "we'll manage" and trying to deflect the conversation to other topics. He brought Helen to a free lunch seminar recently that discussed government entitlement programs that could help with the cost of care.

Overview of financial ability and aging

As individuals age, there can be questions regarding retention of financial capacity. Financial capacity can be defined broadly as the ability to manage one's personal finances in a manner consistent with personal self-interest. Overall, the literature is mixed regarding exactly when financial declines occur, but the overall picture is sobering. Work in cognitive aging suggests that basic cognitive skills such as speed of processing and working memory begin to erode in the third decade of life [1]. However, functional changes

Evidence-informed Approaches for Managing Dementia Transitions
ISBN 978-0-12-817566-8
https://doi.org/10.1016/B978-0-12-817566-8.00004-8

in decision-making and financial skills do not usually emerge until much later, typically in the sixth decade [2]. This is partly due to other aspects of lifespan development such as increased experience and the use of compensatory strategies (on-line banking). The literature regarding financial capacity in persons with dementia is more straightforward. Financial capacity may be comprised very early in the dementia process as it is a complex instrumental activity of daily living [3]. This picture becomes increasingly complicated in caregiver dyads, with one partner aging and a second being a person with dementia. Aging, dementia, psychological vulnerability, and caregiving burden all individually serve as risk factors for financial exploitation. The goal of this chapter is to review the relevant literature and discuss elements related to the balance of supporting financial autonomy for people with dementia with the risks of financial exploitation.

Normal aging and financial decision-making: The literature on normal aging and financial decision-making suggests that while many basic skills of financial decision-making (coin counting, writing a check, bill paying) remain intact until later life, more complex skills (e.g., statement review, obtaining fair loans, investments) may begin to decline [4]. Other research has emphasized that while many older adults demonstrate sound decision-making, there is a subgroup of older adults who demonstrate declines in risk assessment and decision-making secondary to changes in the orbital-frontal cortex [5]. Agarwal et al. examined financial mistakes across the lifespan in healthy individuals [2]. These errors included suboptimal use of credit card balance transfer offers, misestimation of one's house value, and excess interest payments. The authors describe a U-shaped function with optimal financial decision-making occurring at 53 years and increases in errors in older and younger samples. Despite considerable experience, older adults as a group, particularly those older than 80 years, make a greater number of financial errors than middle-aged individuals.

Financial literacy, numeracy, and aging: Financial literacy is a core domain related to financial competence. Financial literacy refers to the basic knowledge and skills required to manage financial resources effectively for a lifetime of financial security. Financial literacy differs from financial capacity in the emphasis on knowledge and skills and not necessarily the individual's particular financial situation. In research studies, typically financial literacy has been assessed using the "Big Three" financial literacy items. These items are:

1. "Suppose you had $100 in a savings account and the interest rate was 2% per year. After 5 years, how much do you think you would have in the account if you left the money to grow?"

A. More than $102
B. Exactly $102
C. Less than $102
D. Don't know
E. Refuse to answer

2. "Imagine that the interest rate on your savings account was 1% per year and inflation was 2% per year. After 1 year, with the money in this account, would you be able to buy ..."
   A. More than today
   B. Exactly the same as today
   C. Less than today
   D. Don't know
   E. Refuse to answer

3. "Do you think the following statement is true or false? Buying a single company stock usually provides a safer return than a stock mutual fund."
   A. True
   B. False
   C. Don't know
   D. Refuse to answer

Using these three items and based on data from the Health and Retirement Survey (HRS), Lusardi and Mitchell have reported that older adults as a group perform more poorly on items related to financial literacy than younger and middle-aged adults [6,7]. Their studies show that financial literacy is related to the ability to correctly answer financial questions, to make better financial decisions about their retirement saving, and to pay off loans and credit cards [7,8]. They also note a gender bias with older men outperforming older women. These effects (age and gender) may compound when older wives become caregivers for older husbands with dementia, resulting in increased vulnerability.

Numeracy is a distinct construct from basic calculation abilities and is related to an individual's comfort with numbers and subsequent use of them in their decision-making [9]. Because older adults tend to be less numerate than their younger counterparts, they may be at higher risk for poor financial decision-making.

In a recent study examining cognitive risk factors for financial exploitation among older adults, high numeracy was found to be a significant predictor of decreased risk after controlling for other demographic variables

[10]. Less numerate participants' self-reported risk was significantly higher as assessed by the Older Adult Financial Exploitation Measure [11]. Importantly, numeracy remained a significant predictor in the presence of other risk factors, including dependency, physical and mental health, as well as overall cognition. Among healthy adults, researchers have shown that higher cognitive abilities are associated with increased likelihood to plan for the future and participate as investors in financial markets [12]. This is particularly relevant given that with an aging population, the prevalence of cognitive disorder is likely to increase.

Protecting one's finances: intergenerational communication and conflict

To date there has been relatively little conceptual and empirical work to help understand the dynamics between older adults and their adult children with regard to finances. In a best-case scenario, the older adult remains independent with her finances until the end of life and there is little or no need to communicate about finances with her adult children. However, with more and more individuals surviving into their late 80s or 90s and beyond, resulting in higher prevalence rates of dementia, frailty, and chronic diseases, the need to communicate about finances becomes more necessary. Losses of community mobility, independence, or financial management skills trigger a new vulnerability related to financial management by older adults. The need therefore to address one's financial situation with adult children becomes more desirable. One theoretical model applied to the communication between the generations around finances is Communication Privacy Management [13].

Plander describes the Communication Privacy Management theory as framed within the tension between both needing to be social and needing autonomy. The three main principles of Communication Privacy Management are (1) Privacy ownership; (2) Use of privacy rules to control access to and protection of information; and (3) Privacy turbulence—how information is used and when its uses violate expectations of the older adult. In Plander's research older adults frequently maintained a high level of privacy about their finances with their adult children. Of note, older adults often viewed disclosure of financial information to their children to be a high risk and low reward communication. In other words, older adults often feared the consequences of revealing the details of their financial information.

Plander utilized an in-depth qualitative interview for 20 dyads of older adults and children in her study of Communication Privacy Management as applied to financial communications between the generations. The findings were telling especially when it comes to why communication about finances are so difficult. Generally, Plander found that there were thick privacy rules and high control goals while both older adult parents remained well. That is, adult children frequently responded that they had no idea what their parents' financial situation was and the older parents reported that they had a high need for controlling their own finances. This high need to control finances and financial information was, for most of these older adults, one of their top goals. Eventually, after a health decline or death of one of the parents, adult children did become involved with the management of their parents' finances. This necessitated a renegotiation of the privacy rules; something that Plander did not describe or expand on.

In the winter and spring of 2018, the Wayne State University caregiver project conducted four focus groups with adult children caregivers to better understand how, when, and why adult children became involved in assisting with financial management for their parents. Our groups were diverse both racially (African Americans and whites) and across socioeconomic status. Across all the groups, we found a consistent piece of information; in nearly 30% of cases, cognitive decline and dementia in an older parent was discovered because of financial exploitation of and/or financial mismanagement by the older parent. Most caregivers cited their observation of their parent's handling of the mail as a trigger for discovering the exploitation. Often there was a proliferation of "lottery" or "sweepstakes" mailings accompanied by a high volume of checks sent out by the older adult to pay the "fees" for their winnings. In some cases, there were repeated donations to charities, who upon receiving one donation sent out a thank you and a request for a new one. Owing to the problems of memory dysfunction, the older adult lost track and ended up sending a new check. In other cases, the adult children noted the high volume of solicitation calls and conversations between their parents and these callers.

The combination of high privacy rules and cognitive and psychological vulnerabilities make older adults vulnerable

Lichtenberg et al. described quantitative findings of a comparison of community members who suffered suspected financial exploitation during the previous 2 years versus those who had not suffered any financial exploitation [14]. The results indicated that those who had experienced exploitation reported significantly greater psychological vulnerability and susceptibility on the self-report items of the Lichtenberg Financial Decision Rating Scale. These studies also provided preliminary interrater reliability and convergent validity for the Lichtenberg Financial Decision-making Rating Scale [15,16]. More recently, in a sample of 200 independent community-dwelling adults, we used factor analysis to test its conceptual model and found support for three contextual factors and an intellectual factor [17]. The contextual factors assess issues related to financial awareness (e.g., self-efficacy, strain, and knowledge); psychological vulnerability with regard to finances; and susceptibility to influence or exploitation due to social vulnerabilities. We used the intellectual factors identified 30 years ago by Appelbaum and Grisso: communication of choice, understanding, appreciation, and reasoning [18]. The Lichtenberg Financial Decision-making Rating Scale is unique in yielding a quantified risk score which is related to both financial management skills and neurocognitive variables. The psychological vulnerability section of the scale explores psychological and cognitive symptoms related to financial decision-making (i.e., anxiety about a transaction). Despite sampling across socioeconomic status, 60% of our sample indicated that they were worried about having enough money to pay for things, although slightly over half reported being satisfied with their financial situation. Nearly 30% reported lacking confidence in making big financial decisions, and nearly 30% reported feeling worried about financial decisions they recently made. Taken together, we can see that in this sample of 200 independent living community-dwelling older adults, there was a significant amount of anxiety regarding financial decision-making as well as relatively low financial decision-making confidence. Finally, nearly half of the sample sought advice about their finances; nearly two-thirds of these individuals stated that this was different from how they handled their finances when they were younger.

In addition to psychological vulnerabilities, participants also indicated declines in social networks that negatively impacted their financial management. Forty-one percent of the sample stated that they wished they had someone to talk to about their finances. Nearly a third of the sample had lost a confidante within the past 2 years, although 81% still stated that they had one. Nearly 40% of the sample stated that they felt anxious about their finances, while slightly over 40% reported that they felt downhearted and blue about their finances at least some of the time. Over a third of the sample stated they were treated with less respect than others during a financial transaction, and one in seven was worried about losing their financial freedom. Finally, nearly a third of the sample reported that their relationship with a loved one was strained because of finances, as they have grown older; with 70% of those stating that the strain has seriously damaged their relationship with that loved one. In sum, there was considerable psychological vulnerability among the older adults in our sample. Anxiety and depressive symptoms related to finances, combined with feelings of low status, few connections and high relationship strain can all impact the privacy rules older adults set around their communication about finances.

The vulnerability of older adults around finances may indeed make it more likely that they are more susceptible to scams and fraud. Although older adults may resist talking with their adult children around finances, the pressures of economic insecurity, psychological vulnerability, and relationship strain around money may impact older adults' desire and willingness to talk with others about their finances. The strong privacy rules that Plander described, ironically, may lead older adults to share information about finances with complete strangers while maintaining strict boundaries with their adult children. These are hypotheses that need further research and evaluation.

Older adult vulnerability to financial exploitation

No one is immune to being financially exploited, and older adults overall do not seem more predisposed to exploitation than any other group, with three exceptions: those with dementia, those with psychological vulnerabilities, and those with financial decision-making deficits. The links between impaired financial decision-making and financial exploitation have been explored in three major ways. First, the impact of neurocognitive disorders on financial capacity across a variety of domains (including decision-making) has been examined [19], and clear evidence has been found for the profound impact of dementia on financial capacity (and by extension vulnerability to financial exploitation). Second, financial decision-making has been studied more explicitly; its links to neurocognitive tests and brain functioning have been detailed, along with the relationships between decision-making and a measure of scam susceptibility [19–21]. Third, more recently, the integration of psychological variables with cognitive variables has been described [15,17,19]. Spreng, Karlawish, and Marson draw on normative aging and decision-making research to highlight the potential role of trust, positivity bias, and deception in financial exploitation [19]. The biggest change in this model is the incorporation of psychological and social phenomena from the normal aging literature. Their new model highlights the roles of social, psychological, neural, and financial management abilities in both decision-making and exploitation risk. Thus, they argue that as a consequence of normal aging, older adults are more trusting, more easily deceived, and less attentive to the potential negative characteristics of others, all of which lead to a higher risk of exploitation.

Two studies have examined the effects of psychological vulnerability and susceptibility in cases of suspected financial exploitation [16,22] and nonnormative aging experiences (e.g., anxiety, depression, low sense of social status, low sense of financial mastery and reduced security). These found that more extreme psychological vulnerability led to higher rates of fraud victimization, in both cross-sectional and longitudinal samples.

Financial decision-making deficits are linked to financial exploitation in many cases involving older adults. In a recent case, for example, a man with early dementia was taken to the bank by his brother, and before he left the bank, he paid the remainder of his brother's home mortgage of more than $100,000. In another case a successful retired businessman with executive dysfunction difficulties, alone and lonely, lost nearly $1 million in a scam in which he was convinced that a woman he had never met loved him, and needed to marry him and take him to Italy to receive a $20 million inheritance. These two cases illustrate how financial decision-making and financial exploitation are linked. Until recently there were no efficient ways to measure financial decision-making and associated risk.

Enhancing communication between adult children and older parents around finances

One clear implication of Plander's findings, especially the belief of many older adults that risks outweigh benefits in disclosing financial information to their adult children, is that communication about finances and financial safety between older adults and their adult children is extremely difficult. Combine that with the increasing numbers of older adults who become more vulnerable to financial exploitation and the need for new approaches is evident. Too often, when adult children are worried about their parents, and particularly when the worry is about financial exploitation, the children move aggressively to take over or convince the older adult that what they are doing is wrong and even possibly dangerous. This often backfires.

These conversations are exactly the type of interactions that Patterson et al. term *Crucial Conversations* in their book carrying that title; conversations in which strong emotions, differing opinions, and high stakes are involved [23]. Patterson et al. emphasized that psychological safety is imperative during crucial conversations [23]. Without psychological safety, the conversations may lead to poor outcomes and, in the case of vulnerable older adults, maybe even lead to more susceptibility to financial exploitation. Patterson et al. noted that silence/withdrawal or anger/outbursts are two common indicators that the conversation is feeling unsafe [23]. How, then, to best enhance safety in conversation about finances? Other work on difficult conversations applies here.

The work of Stone et al underscores the hidden elements of difficult conversations and how to effectively negotiate them [24]. First, they note that there are three levels of conversation: (1) the "what happened" conversation; (2) the "feeling" conversation; and (3) the "identity" conversation. Most conversations focus on what happened: "just the facts." The utility of the conversation, however, depends more on attention to the feelings involved and how the conversation reflects on an individual's identity as competent, caring, a good son, father, daughter, or mother. The authors note that, so often, a number of assumptions are made (e.g., that we know all the facts we need to) that derail conversations. Abandoning erroneous assumptions and seeking to partner with the older adult can produce better results in communication.

The goals of these difficult conversations are to help older adults improve their financial safety. Some of the guiding principles for effective partnering with older adults around financial safety and health involve principles found in motivational interviewing. The goal of motivational interviewing is to strengthen a person's motivation and commitment to behavior change by helping them resolve their ambivalence about the change. Principles of this approach include (1) Partnership: staying outside the expert or advice giving role; (2) Acceptance and avoiding confrontation; (3) Compassion: being empathetic and an advocate for the older adult; and (4) Evocation: eliciting the older person's reasons for change, expressing curiosity.

In this spirit we developed a set of open-ended questions that allow for nonthreatening conversation about finances and financial health and safety. These can be used as part of a clinical intervention or alternatively as nonthreatening techniques to start a difficult conversation within one's family. Some of these include

- I have been learning that older adults are increasingly the targets for financial exploitation; what do you know about financial exploitation of older adults?
- Do you know anyone who has been financially exploited?
- Have you ever before been financially exploited?
- Have you noticed any changes in your financial management skills as you have gotten older?
- Any changes in the way you make financial decisions such as seeking more input and advice than you used to?
- Have you thought about ways to protect your finances?
- What is it that I don't know or understand about the ways you protect your finances now?
- Is there a possibility that adding ways to protect your finances would be a good idea?
- Should you consider adding ways now when you are well or wait until things become problems?
- Under what circumstances would you like me to know about your finances? What would you like me to do at that time?

It is very important throughout to affirm the older adult's strengths and goals and to reflect on and deeply listen to what they are saying. Being able to summarize what they have said is a good way to deepen and focus the conversation.

Cognitive impairment and financial decision-making in older adults

The dramatically increasing numbers of older adults in America underscores the fact that the number of people with cognitive impairment will close to triple in the next 35 years [25]. Overall, decline in cognition is associated with a decrease in financial competency in older adults [26]. Data from the Rush Memory and Aging Project indicated that declines in episodic memory and visual spatial ability have been linked to susceptibility to scams [20]. Declines in cognition also predicted a drop in self-confidence in general, but not with a drop in confidence regarding the management of one's own finances [26]. This discrepancy between "confidence" and "competence" can create a number of challenges. For example, it may become a source of tension between older adults and their adult children who can see the erosion in skills but who resist positive intervention. Furthermore, the lack of insight can create a vulnerability that can be exploited by unethical individuals. Based on the Rush data, older adults experiencing decreases in cognition do show an increased likelihood of getting help with financial decisions; however, many participants experiencing significant drops in cognition still do not get help.

Financial ability and dementia

The impact of age-related dementia (e.g., Alzheimer's disease) on financial capacity threatens financial autonomy [27,28]. Marson et al have examined how aging and major neurocognitive disorders impact financial capacity, defined by them as the ability to manage money and financial assets in ways consistent with one's values or self-interest. It is now understood that dementia syndromes may have an onset that lasts decades with relatively "mild" symptoms emerging years before a full-blown dementia syndrome [29,30]. In other cases, the "mild" symptoms do not progress but persist resulting in subtle declines. Mild cognitive impairment refers to the stage of the illness where cognitive deficits are present, but day-to-day functioning is relatively intact. The pattern of deficits varies by etiology, but many individuals manifest deficits in memory, executive functioning, and calculation during this stage of the disease. Han et al reported that in a sample of older individuals without dementia, decreased gray matter volume in frontal and temporal regions was significantly related to susceptibility to telemarketing scams [31]. In many ways, those individuals with predementia and mild cognitive impairment are the perfect victim. They retain control of their assets and are out in the community increasing their exposure. As reviewed previously, these individuals have little insight into their vulnerabilities and retain formal barriers in discussions with caring family members.

Loss of financial autonomy during dementia: findings from qualitative research

Knowing that our worldwide population is aging, that the prevalence of dementia is greater in older age groups, and that many older adults (1) may not communicate their financial situation to adult children and (2) are not confident in their abilities to manage money the older they get, persons with dementia and their families need strategies to reduce the level of financial vulnerability. The following sections draw on the results of the Dementia Transitions Study described in Chapter 1.

The study sought to determine the impact of environmental, functional, and personal factors on the overall well-being of individuals with dementia during key periods of transition across the illness timeline. The data presented here focus on a subset of questions relating to loss of financial autonomy. Financial stressors impact more than the individual with the dementia. For instance, Harary et al measured the impact of dementia care on caregivers' professional, personal, emotional, and social well-being in a cohort study of 1387 caregivers [32]. They found that 25% of participants reported having to reduce their work schedules, and many participants reported more emotional and financial stress. Of the 108 participants who took part in the Dementia Transitions Study, 32 experienced loss of financial autonomy.

Questions from the study were designed to identify and evaluate the person with dementia's and caregiver's perspectives of the effectiveness of financial assistance programs or other methods used to reduce the burden of stress associated with the transition. Having open-ended questions as a guide for the interview process allowed the people with dementia or their caregivers an opportunity to raise any issues they believed to be important regarding their experiences with a loss of financial autonomy. In addition to the generic questions outlined in Chapter 1, the following questions explored the impact of dementia on financial autonomy from the perspective of the caregiver:

- Which of the person's financial affairs were becoming problems?
- What methods have you adopted to assume control over their financial affairs?
- Do they know about and agree to assuming financial control, and if not, what procedures, processes, or tactics do you use to achieve your aim without their knowledge and/or agreement?

The interview guide for the transition to loss of financial autonomy included the following question for the person with dementia:

- What have you found particularly difficult about having your caregiver look after your finances?

The initial process of coding began by reading through the entire interview transcripts and familiarizing oneself with the data. Once the major patterns in the transcript data had been identified, a preliminary coding process was developed with loss of financial autonomy in mind. The analysis and coding was conducted through the use of NVIVO software, and the information was kept confidential throughout the entire process. Using content analysis, the data were reviewed and a preliminary list of 53 relevant codes were defined. These were then regrouped into higher order groups by two additional coders. The resulting 19 codes are listed below.

From the coded data emerged three major themes. A "sign" indicated a change in financial autonomy without evidence in the transcript of any positive or negative impact. A "challenge" involved a struggle or obstacle that must be overcome in relation to financial autonomy, and a "strategy" was defined as a method used to reduce the burden of stress on the person with dementia or caregiver in relation to financial autonomy.

Codes

1. Signs:
   - **A.** Person with dementia misplacing money, bank cards, or bills.
   - **B.** Person with dementia losing interest in finances.
   - **C.** Incurring debt due to poor financial management skills.

2. Challenges:
   - **A.** Person with dementia navigating paperwork.
   - **B.** Caregiver feeling financial burden in addition to care burden.
   - **C.** Signing legal papers while living with dementia.
   - **D.** Caregiver having difficulty invoking Power of Attorney.
   - **E.** Person with dementia and caregiver experiencing misinformation about financial aid.
   - **F.** Person with dementia being financially abused by others.

3. Strategies:
   - **A.** Person with dementia maintaining some control over some of the finances.
   - **B.** Caregiver taking over all finances.
   - **C.** Person with dementia maintaining some real control of their finances.
   - **D.** Caregiver assuming Power of Attorney and personal directive.
   - **E.** Person with dementia choosing someone else to manage money.
   - **F.** Person with dementia maintaining some perceived control of their finances.
   - **G.** Person with dementia or caregiver implementing direct deposit for bills.
   - **H.** Person with dementia maintaining use of bank card by utilizing PIN number.

I. Power of Attorney being set up before dementia as a precaution.
J. Caregiver or person with dementia using financial aid to decrease financial burden.

Signs, challenges, and strategies

It was clear from the transcripts that caregivers identified "signs" of reduced financial competence early on but did not see any real impact or consequence on functioning. The most common sign cited in the interviews was the person with dementia misplacing money, bank cards, or bills, often followed by the person with dementia losing interest in finances. These signs were sometimes dismissed by the caregiver, and no real action was taken until the lack of financial autonomy became worse.

Caregiver: Like I say, the only thing I would say is losing her bank card.

Interviewer: Was she upset?

Caregiver: Well she noticed because she didn't have it but not overly concerned. But I think that is kind of the progression. She is becoming less and less sort of engaged with these issues.

Misuse or loss of a bank card can be one of the earliest signs of difficulties with finances, but these can also be seen as just moments of forgetfulness. Giebel et al identified paying bills and managing taxes as the two functional activities that were the most affected early on in dementia [33]. Variations in impairment were found depending on the type of dementia. The authors stress the importance of attending to these early signs to prevent further impact on financial vulnerability. Misuse of bank cards is a common risk factor for financial exploitation.

Caregiver: It is just a matter of getting the bill in hand. I got her hydro one just the other day. Problem I have now is her tax statement stuff is coming in and I do her taxes every year. Normally she takes it all and puts it in one pile, but she can't seem to manage that anymore. She is putting it away, hiding it, opening it up. She has really lost all ability to manage paper work. It's hit and miss, sometimes she is there and sometimes she is not.

For some, as in the aforementioned example, the impact is financial, but the cause is ancillary to the function of financial competence (i.e., managing paperwork in this case). Caregivers certainly became more concerned when debt was incurred because of poor financial management skills. The impact was greatest in cases where the person with dementia had previously been responsible for all the finances. In these cases, the caregiver may never have thought to check the bank accounts or bills until it was too late and debt was incurred. In these situations, both partners may feel overwhelmed but retain high communication boundaries. The most common strategy used to effectively manage the challenge of misplacing bank cards or bills was to either have the caregiver take over all the finances or to establish a direct payment system from the bank account. Although the first could alleviate some of the stress experience by the caregiver, the burden of financial responsibility did add to the overall burden already experienced as a result of other responsibilities. Having their finances taken over by another has the added effect of reducing the person with dementia's sense of financial autonomy and independence and could lead to negative emotional reactions.

The desire to allow the person with dementia to maintain autonomy and independence is contrasted with the desire to not incur debt. As mentioned previously in this chapter, the best prevention would be careful advance planning and sharing of information predementia so that the impact is lessened. For the Dementia Transitions Study participants, the most common strategy implemented did indeed involve allowing the person with dementia some control over some of the finances. Allowing people some cash in their wallets, for instance, recognizes their sense of independence and reduces anxiety in the event of an emergency should the person with dementia wander or become lost. Similarly, caregivers found that persons with dementia could even manage a bank card, provided some restrictions were placed on PIN numbers and amounts that could be withdrawn.

Interviewer: Do you use a credit card and debit card?

Person with dementia: Yeah, my credit cards all have PIN numbers and very simple PIN numbers and it's the same for all credit cards.

Interviewer: And does that help you by having a PIN number?

Person with dementia: Yes. Because I can't, I don't have the ability to sign my name anymore.

Some caregivers assume that the person with dementia is unable to recognize that there is a need to adapt to his or her growing functional limitations and may be inclined to withdraw all control of finances from them. In fact, in the early stages of dementia, persons with dementia may recognize their deficits in doing complex financial tasks while not recognizing the impact on simpler functions [34]. This may suggest that persons with dementia may be more open to developing new strategies to adapt to these challenges. Caregivers would benefit by communicating openly with persons with dementia about the limitations while allowing the person some control. This would smooth the transition to eventual total loss of financial autonomy. Caregivers do agree that letting the person with dementia maintain some control over some of their finances using bank cards or cash proved to be a useful strategy in reducing anxiety and financial stress experienced by the caregiver.

Interviewer: Good, and finances. How is, is he handling finances? He still has money?

Caregiver: He still has money. Um, he still thinks he has to pay when we go to restaurant, which is fine. Um, we go to buy gas, he knows he has to pay

Interviewer: And does he use a debit card or credit card or cash?

Caregiver: He cannot use a debit card. We tried that in the [name of restaurant]. And once the waiter gave him the machine, he has no idea and passed it over to me.

Interviewer: Okay, so that he feels like he is paying but you are doing the transaction.

Caregiver: Yes

Interviewer: And um, he goes to the bank and gets cash out ... does he use the debit card at the bank machine?

Caregiver: He uses the, [name of bank] convenience card.

By not engaging in a struggle about finances, the person with dementia may come to realize his or her difficulties and even themselves choose to delegate the financial responsibilities to their caregiver. This allows the person with dementia to maintain their self-esteem and reduced the caregiver's uncertainty about whether the person with dementia would want them to handle financial affairs. Letting the person with dementia come to that realization while keeping an eye on the finances might be the best strategy to use.

Interviewer: I just want to check in and see how everything is doing, how your mom is doing and everything.

Caregiver: Ah pretty good actually, we were just having a conversation and she's finally decided to let me take over all the financials.

Interviewer: Oh that's great.

Caregiver: Phew, tell me it's a huge load off my mind and it's like I said, that's the only time when we have arguments is when she's not remembering when she's getting her statements and she's freaking out, so she finally just said today, she says, 'you know what, you do everything, I don't want nothing to do with that'. So that'll, that makes my life easier.

In fact, in many cases a legal transfer of power in the form of a *Power of Attorney* which legally allows the caregiver to make financial decisions on behalf of the person with dementia is an effective method to handle some of the challenges associated with transitioning to complete loss of financial autonomy. For participants in the Dementia Transitions Study, this strategy was effective because the person with dementia was aware that they had delegated financial control to their caregiver and had faith in their decision-making ability.

In fact, dyads that experienced the least challenges and struggles were the ones who decided to set up Power of Attorney arrangements before the dementia progression, as a precaution. Creating a Power of Attorney agreement was one of the most frequently cited strategies by participants. They agree that individuals experiencing early stages of dementia would likely benefit from speaking with their financial advisors and delegating an individual to act on their behalf. A relatively simple step like this can greatly reduce confusion, anxiety, and legal issues involving finances in the future. There is evidently a need to inform the public of the importance of clearly identifying an individual to act on one's financial behalf should the need ever arise. Considering that this strategy was also associated with the fewest number of challenges, it provides some support for the need to have a financial action plan prior to the complete loss of financial autonomy.

Conclusion

One of the most cited results from the Dementia Transitions Study is consistent with the findings of the study by Martin et al in 2013 that suggested individuals with mild cognitive impairment transitioning to dementia often experience a gradual decline in financial capacity [35]. A review by Giebel et al in 2015 examined 17 studies on everyday functioning and 40 studies on memory to find that complex instrumental activities of daily living were subject to greater impairment than basic activities of daily living [36]. They found that individuals who were in the early stages of dementia struggled with financial tasks; a deficit linked to impaired working memory. A challenge that was commonly cited by participants in the Dementia Transitions Study was that the caregiver experienced misinformation about financial aid and had issues invoking a Power of Attorney or signing legal papers. These challenges are confirmed by Vaingankar in 2013, who identified lack of financial support and information as a key unmet need for caregivers [37].

Some key strategies that proved to be effective for persons with dementia and caregivers during this transition involved the person with dementia maintaining some control over a portion of the finances. From the interview transcripts, this strategy appeared to be far more effective than having the caregiver take over all finances. This may be because the former strategy allowed the person with dementia to maintain some sense of autonomy and independence with regard to their lives. It is well established that caregivers are under increased psychological burden and have fewer resources to cope than noncaregivers. Identifying the transition to a caregiving role and intervening with informal or formal supports becomes even more important in these situations.

More research is needed to identify the most effective strategies for transitioning to loss of financial autonomy and in which contexts. Results showed some very innovative and useful strategies that were used by caregivers to reduce the stress associated with the transition. Some of these strategies include setting up a direct deposit to ensure bills were paid on time and implementing a Power of Attorney before the transition.

Since most caregivers who were helping persons with dementia during this study were either spouses or children, it may be beneficial to introduce awareness campaigns targeted toward the baby boom population as well as older adults in the community.

Unfortunately, one theme from the study also pointed to the possibility of financial abuse and exploitation. This should be taken very seriously considering the large number of people currently living with dementia and the predicted number in the future. This is supported by current literature through a study conducted by Byrne et al in 2015 about financial capacity in older adults [38]. They summarize "Impairment of financial capacity makes the older individual vulnerable to financial exploitation, may negatively impact their family's financial situation and place strain on relationships within the family"(p. 82). To prevent this strain and exploitation in the future, it may be beneficial to investigate financial abuse in individuals with dementia and develop guidelines for prevention.

This chapter sheds light on some common signs to watch out for and the challenges one is likely to expect. Early intervention would allow for smoother transfer of financial responsibilities and perhaps avoidance of distress on the part of those with declining financial capacity and their caregivers. For a society to maintain a good quality of life for its populations, it must seek to better understand how loss of financial capacity can impact persons with dementia physically, socially, and psychologically. The goal of this chapter was to focus clinical and professional attention on a neglected capacity that is of fundamental importance for patients, families, and the healthcare system. Being aware of common signs, challenges, and strategies will undoubtedly better prepare caregivers and persons with dementia to navigate the transition effectively.

References

[1] Salthouse T. Selective review of cognitive aging. J Int Neuropyschol Soc 2010;16(5):754−60. https://doi.org/10.1017/S1355617710000706.
[2] Agarwal S, Driscoll JC, Gabaix X, Laibson D. The age of reason: financial decisions over the life-cycle and implications for regulation. Brook Pap Econ Act 2009;2:51−117. http://muse.jhu.edu/article/372531.
[3] Griffith HR, Belue K, Sicola A, Krzywanski S, Zamrini E, Harrell L, Marson DC. Impaired financial abilities in mild cognitive impairment: a direct assessment approach. Neurol 2003;60:449−57.
[4] Bangma DF, Fuermaier ABM, Tucha L, Tucha O, Koerts J. The effects of normal aging on multiple aspects of financial decision-making. PLoS One 2017;12(8):e0182620. https://doi.org/10.1371/journal.pone.0182620.
[5] Denburg NL, Cole CA, Hernandez M, Yamada TH, Tranel D, Bechara A, et al. The orbitofrontal cortex, real-world decision-making, and normal aging. Ann N Y Acad Sci 2007;1121(1). https://doi.org/10.1196/annals.1401.031.
[6] Lusardi A, Mitchell OS. Financial literacy and retirement preparedness: evidence and implication for financial education. Bus Econ 2007;42(1):35−44.
[7] Lusardi A, Mitchell OS. Financial literacy around the world: an overview. NBER Work Pap Ser 2011. http://www.nber.org/papers/w17107.
[8] Lusardi A, Tufano P. Debt literacy, financial experience and overindebtedness. NBER Work Pap Ser 2009. http://www.nber.org/papers/w14808.
[9] Reyna VF, Nelson WL, Han PK, Dieckmann NF. How numeracy influences risk comprehension and medical decision making. Psychol Bull 2009;135(6):943−73. http://doi.org/10.1037/a0017327.
[10] Wood SA, Liu PJ, Hanoch Y, Estevez-Cores S. Importance of numeracy as a risk factor for elder financial exploitation in a community sample. J Gerontol B Psychol Sci Soc Sci 2015;71(6):978−86. http://doi.org/10.1093/geronb/gbv04.
[11] Conrad KJ, Iris M, Ridings JW, Langley K, Wilber KH. Self-report measure of financial exploitation of older adults. Gerontologist 2010;50(6):758−73. https://doi.org/10.1093/geront/gnq054.
[12] Cole SA, Shastry GK. Smart money: the effect of education, cognitive ability, and financial literacy on financial market participation. Work Pap 2009. http://www.people.hbs.edu/scole/webfiles/smarts-revised-08-02.pdf.

[13] Plander KL. Checking accounts: communication privacy management in familial financial caregiving. J Fam Commun 2013;13:17–31.
[14] Lichtenberg PA, Ocepek-Welikson K, Ficker LJ, Gross E, Rahman-Filipiak A, Teresi JA. Conceptual and empirical approaches to financial decision-making by old adults: resultsfrom a financial decision-making rating scale. Clin Gerontol 2018;41(1):42–65. https://doi.org/10.1080/07317115.2017.1367748.
[15] Lichtenberg PA, Stoltman J, Ficker LJ, Iris M, Mast BT. A person-centered approach to financial capacity assessment: preliminary development of a new rating scale. Clin Gerontol 2015;38:49–67. https://doi.org/10.1080/07317115.2014.970318 JGP. 0b013e318157cb00.
[16] Lichtenberg PA, Ficker L, Rahman-Filipiak A, Tatro R, Farrell C, Speir JJ, et al. The Lichtenberg financial decision screening scale: a new tool for assessing financial decision making and preventing financial exploitation. J Elder Abuse Negl 2016;28:134–51. https://doi.org/10.1080/08946566.2016.116833.
[17] Lichtenberg PA, Ocepek-Welikson K, Ficker LJ, Gross E, Rahman-Filipiak A, Teresi J. Conceptual and empirical approaches to financial decision making in older adults: results from a financial decision-making rating scale. Clin Gerontol 2017;41:42–65. https://doi.org/10.1080/07317115.2017.1367748PMC5766370.
[18] Appelbaum PS, Grisso T. Assessing patients' capacities to consent to treatment. N Engl J Med 1988;319:1635–8. https://doi.org/10.1056/NEJM198812223192504.
[19] Spreng RN, Karlawish J, Marson DC. Cognitive, social and neural determinants of diminished decision-making and financial exploitation risk in aging and dementia: a review and new model. J Elder Abuse Negl 2016;28:320–44. https://doi.org/10.1080/o8946566.2016.1237918.
[20] Boyle PA, Wilson RS, Yu LY, Buchman AS, Bennett DA. Poor decision making is a consequence of cognitive decline among older persons without Alzheimer's disease or mild cognitive impairment. PLoS One 2012;7(8):1–5. https://doi.org/10.1371/journal.pone.0043647.
[21] Han S, Boyle PA, Yu L, Arfanakis K, James BD, Fleischman DA, Bennett DA. Grey matter correlates of susceptibility to scams in community-dwelling older adults. Brain Imaging Behav 2015;10(2):521–32. http://doi.org/10.1007/s11682-015-9422-4.
[22] Lichtenberg PA, Stickney L, Paulson D. Is psychological vulnerability related to the experience of fraud in older adults? Clin Gerontol 2013;36:132–46. https://doi.org/10.1080/07317115.2012.749323.
[23] Patterson K, Grenny J, McMillan R, Switzler A. Crucial conversations. New York: McGraw Hill; 2012.
[24] Stone D, Patton B, Heen S. Difficult conversations. London: Penguin Books; 2010.
[25] Hebert LE, Scherr PA, Bienias JL, Bennett DA, Evans DA. Alzheimer disease in the US population: prevalence estimates using the 2000 census. Arch Neurol 2003;60:1119–22. https://doi.org/10.1001/archneur.60.8.1119.
[26] Gamble P, Boyle P, Yu L, Bennett D. Aging and financial decision-making. Manag Sci 2015;6(11):2603–10.
[27] Marson DC. Loss of financial competency in dementia: conceptual and empirical approaches. Aging Neuropsychol Cognit 2001;8(3):164–81. http://doi.org/10.1076/anec.8.3.164.827.
[28] Griffith HR, Belue K, Sicola S, Krzywanski E, Zamrini E, Harrell L, Marson DC. Impaired financial abilities in mild cognitive impairment. Neurol Now 2003;60(3). Error! Hyperlink reference not valid, https://doi.org/10.1212/WNL.60.3.449.
[29] Marson DC. Clinical and ethical aspects of financial capacity in dementia: a commentary. Am J Geriatr Psychiatry 2013;21:392–400.

[30] Martin R, Griffith HR, Belue K, Harrell L, Zamrini E, Marson D. Declining financial capacity in patients with mild Alzheimer disease: a one-year longitudinal study. Am J Geriatr Psychiatry 2008;16:209–19.
[31] Han SD, Boyle PA, James BD, Yu LY, Barnes LL, Bennett DA. Discrepancies between cognition and decision making in older adults. Aging Clin Exp Res 2015;28:99–108. https://doi.org/10.1007/s40520-015-0375-7.
[32] Tang B, Harary E, Kuzman R, Mould-Quevado JF, Pan S, Yang J, et al. Clinical characterization and the caregiver burden of dementia in China. Value Health Reg Issues 2013;2(1):118–26. https://doi.org/10.1016/j.vhri.2013.02.010.
[33] Giebel CM, Flanagan E, Sutcliffe C. Predictors of finance management in dementia: managing bills and taxes matters. Int Psychogeriatr 2019;31(2):277–86. https://doi.org/10.1017/S1041610218000820.
[34] Van Wielingen LE, Tuokko HA, Cramer K, Mateer CA, Hultsch DF. Awareness of financial skills in dementia. Aging Ment Health 2004;8(4):374–80.
[35] Martin RC, Triebel KL, Kennedy RE, Nicholas AP, Watts RL, Stover N, et al. Impaired financial abilities in Parkinson's disease patients with mild cognitive impairment and dementia. Park Relat Disord 2013;19(11):986–90.
[36] Giebel CM, Challis D, Montaldi D. Understanding the cognitive underpinnings of functional impairments in early dementia: a review. Aging Ment Health 2015;19:859–75.
[37] Vaingankar JA, Subramaniam M, Picco L, Eng GK, Shafie S, Sambasivam R, et al. Perceived unmet needs of informal caregivers of people with dementia in Singapore. Int Psychogeriatr 2013;25:1605–19. https://doi.org/10.1017/S1041610213001051.
[38] Gardiner PA, Byrne GJ, Mitchell LK, Pachana NA. Financial capacity in older adults: a growing concern for clinicians. Med J Aust 2015;202:82–5.

CHAPTER 5

Hospitalization of persons with dementia

Katherine S. McGilton[1,2], Geneviève Lemay[3,4]
[1]Research, Toronto Rehabilitation Institute- UHN, Toronto, ON, Canada; [2]Lawrence S Bloomberg, Faculty of Nursing, University of Toronto, Toronto, ON, Canada; [3]Faculty of Medicine, Division of Geriatrics, University of Ottawa, The Ottawa Hospital, Ottawa, ON, Canada; [4]Montfort Hospital/ Institut du Savoir Montfort Ottawa, ON, Canada

Persons with dementia face unique challenges related to management of their general health, as well as difficulties in managing chronic conditions such as hypertension, diabetes, coronary artery disease, heart failure, and chronic obstructive lung disease [1]. More than 90% of community-dwelling persons with dementia are living with two or more coexisting chronic medical conditions [2,3], which increase their risk of acute hospital admission. The challenges they experience in managing their conditions are often due to deficits in memory, insight, judgment, language, and decision-making ability. These challenges can lead to health status deterioration and hospital readmissions, a pattern referred to as the "domino effect" [4]. This chapter describes the prevalence of dementia in hospitalized patients, reasons for hospitalizations, and risks to older people with dementia in hospitals. Although there are significant risks when hospitalized, these threats can be decreased. A description of strategies for optimizing outcomes and models of care that are designed to improve care of persons with dementia will be presented toward the end of the chapter. Quotes from care partners of persons with dementia interviewed about the experience of hospitalization as part of the Dementia Transitions Study described earlier in the book (Chapter 1) illustrate some of the challenges experienced by persons with dementia and their care partners.

Prevalence of dementia in hospitalized patients

The prevalence of dementia across acute care hospital wards varies widely and depends on the demographic profile of patients and the type of ward [5]. Typically, persons with dementia comprise between 20% and 42% of all hospital admissions [6–8]. During hospitalization, persons with dementia are at increased risk of delayed discharge because of the potential for episodes of delirium, functional decline in response to the acute illness, and

Evidence-informed Approaches for Managing Dementia Transitions
ISBN 978-0-12-817566-8
https://doi.org/10.1016/B978-0-12-817566-8.00005-X

other iatrogenic complications. In addition, cognitive impairment can often be undetected before admission to an acute care hospital. In some cases, dementia is first discovered after admission. For example, 15.8% of patients admitted to an orthopedic hospital for elective or traumatic surgery had evidence of significant cognitive impairment without an established diagnosis [9].

For people with dementia who experience challenges such as impaired memory or insight [10], their care in hospital settings represents a challenge for healthcare team members. Issues are encountered when acquiring a medical history and symptomatology from persons with dementia, providing medical recommendations and encouraging self-care. Care provision can be further challenged when those with dementia exhibit behaviors such as agitation or aggression [11]. Although people with dementia constitute a significant number of hospital admissions [8], there is a little progress in staff training and minimal emphasis on understanding effective models of care for persons with dementia in acute care settings [12].

Repeated hospital admissions and increased length of stay are higher in people with dementia than in those without cognitive problems [5,13]. The median length of hospital stay for persons with dementia was found to be twice as long as those without a diagnosis of dementia [13,14]. Acute care environments impose various negative outcomes and are not the best place for persons with dementia [15]. As one daughter observed,

> *I find that [hospital] deconditioned my dad. [...] not taking him to the bathroom as often and finding him, like, on the side of the bed with his cast almost down to his knee or past his knee ... because he had to go to the bathroom. And he didn't know better to ring.*

Complications of hospitalization for persons with dementia include delirium, loss of independence, institutionalization, and death [16]. These complications exacerbate the economic burden of dementia and increase healthcare needs due to loss of independence [2,17,18]. Higher short-term mortality for hospitalized persons with dementia has been consistently reported in the literature. A meta-analysis reported that the mortality in persons with dementia was significantly higher than in nondementia cases [17]. Furthermore, there is evidence that the admission to hospital of a person with dementia with acute medical illness is a critical event associated with high 6-month mortality rates [18]. Compared with other older patients, persons with dementia are over three times more likely to die during a hospital admission and the risk is even higher for those with the lowest cognitive functioning, who are five times more likely to die [19]. These increased risks remain significant even after controlling for age and severity of acute physical illness. For persons with advanced dementia, 6-month mortality after pneumonia was 53%, compared with 13% for cognitively intact patients, and 55% for older adults with hip fractures, compared with 12% of patients without dementia [18]. Family members have good reason to be concerned to leave their relatives with dementia alone in acute care hospitals.

Reasons for hospitalization

During the course of the disease, many persons with dementia require hospitalization either for acute medical problems, dementia-related complications, or issues related to care partners' burnout [20,21]. Evidence shows that persons with dementia have the greater risk of being admitted because of dehydration, psychiatric crisis, and behavioral disturbances [22], and once they are admitted, these issues continue to be a concern for those with dementia [16,18,23].

Several possible explanations of the relationship between dementia and frequent hospitalization have been hypothesized. First, there are underlying conditions that increase the risk of dementia (e.g., stroke) or that develop in the context of an established dementia diagnosis (e.g., dysphagia, aspiration pneumonia) [1]. Second, the effects of dementia on cognitive functions (e.g., impaired executive function, language, the perception of symptomatology and insight) impair the persons' ability to manage chronic conditions, identify new symptoms, and seek medical attention when needed [1]. Third, the threshold for admission to hospital may be lower in older adults, especially in persons with dementia, because their nervous and metabolic systems are more vulnerable to acute illness and are at an increased risk for complications [24,25]. Finally, caregiver burden may also result in increased acute care admissions [11,26]. In fact, disinhibited behaviors, which occur for some persons with dementia, were found to be a predictor of frustration and embarrassment for the care partners and may lead to admission to a hospital.

Risks for persons with dementia in hospital

There are multiple risks for persons with dementia when they are admitted to the hospital. Some are institutional, while others are related to the person with dementia or to their care partners.

Institutional factors

In-hospital experience from admission to discharge can be disorienting to persons with dementia, as some have limited capability to navigate their surroundings, solve problems, and withstand the noise and pace of the environment [27]. Unfamiliar stressors overwhelm, frustrate, and may frighten persons with dementia. This is especially true in the emergency department. The emergency department rooms are confusing, as they are without windows, lack easy access to washrooms, and do not have enough space for family members. As a result, persons with dementia are at heightened risk of adverse outcomes, with the most common adverse outcome being delirium [28]. Comprehensive geriatric assessments are rarely conducted in the emergency department. Knowledge about dementia care in addition to delirium recognition and management are also often missing. There are often precipitous discharges without a discussion about care needs with family members, as well as a lack of continuity of care with community services to prevent readmissions [29–31] causing more burden on the care partners.

As a daughter of a person with dementia explained:

> *She stayed in the emergency for quite a while. I'm telling these people she has dementia, and you know, she's going to tell you she's fine when she is not. And they all say they understand and they don't ... She was in the emergency for 3 days. Now, emergency is busy. There, she is not a priority, she is not an emergency. She stays on a hard gurney ... no shower, no washing, no nothing, no gown, she was in extreme pain, I go to a dark place every time she goes to some place and I'm not with her.*

Many hospitalized older adults with dementia cannot be discharged once treatments are completed because of insufficient health and social support (e.g., community services, nursing home beds) to meet their postacute care needs. Delayed discharges—episodes where a patient stays in an acute care bed for longer than medically necessary—are a critical challenge for many healthcare systems [32]. Older adult patients who experience delayed discharge are at even greater risk of delirium, falls, and infections. In addition, some hospitals charge patients with prolonged delayed discharge a daily copayment, which may place a substantial financial burden on some [33]. Many persons with dementia experience delayed discharge for a multitude of social reasons leaving them to stay in an acute care bed they no longer require. They are then designated as "alternate level of care" (ALC) patients [34], which has become a very complex acute care resource issue in Canada.

Although we know a great deal about the characteristics of patients who end up spending a long time in acute care hospitals, very little is known about what life is like for persons with dementia (PWD) and their families once they are designated as ALC. Investigators in one study focused on delayed hospital discharges found that care partners reported a range of concerns [33]. For example, once designated as ALC, the healthcare staff may overlook the person because they perceive they do not have further medical needs. In addition, care partners expressed uncertainty and confusion while waiting for the long-term care bed. There also needs to be consideration of care required for the care partners as they also have needs and they tend not to be woven into the fabric of health systems, which predominantly cater to the needs of the patient [33]. As Pringle surmised in 1998 [35], an acute care hospital was not designed to be anybody's home and being an ALC patient is like being in the black hole of hospitals. Some members of the healthcare team are angry at you for occupying a bed for which they have other patients with acute care needs, and staff on acute care floors are busy and taking the time to help old, very disabled patients do as much for themselves as possible is time consuming. Unfortunately, not much has changed over the past 20 years.

Another risk related to the focus on efficacy in hospitals is the fact that healthcare professionals in acute care settings do not always adopt attitudes toward older patients, especially those with dementia, that are conducive to a person-centered approach to care [36,37]. Some factors that may impede adoption of person-centered care approaches include the need for urgent patient flow in and out of the emergency room, speedy investigations of the presenting illness, rapid diagnoses, and pressure to initiate discharge from hospital as early as possible to reduce the length of stay. Healthcare professionals' preconceived thoughts about persons with dementia need to be taken into account, as it is likely that ageism plays a role in their approach to patients with dementia. Negative attitudes regarding older adults persist in our society [38]. It is possible that healthcare professionals' preexisting stereotypes about age and/or dementia will lead them to reinforce dependency. An ageist approach of disregard and lack of recognition would not be conducive to patient-centered care. At the hospital ward level, healthcare team members are also pressured by unit cultures [39]. Furthermore, stereotypes are in play at systemic levels. For example, rehabilitation for older adults with dementia is often denied because of beliefs that people with dementia cannot be rehabilitated, even though this myth has been dispelled [40]. As a consequence, these older adults frequently become designated as ALC patients.

There is often a lack of communication with the person with dementia and care partners from the treating practitioner and entire healthcare team related to not only the preexisting diagnosis of dementia but also the current diagnosis and treatment plan. The patient's and care partner's interaction with staff in an acute care setting is dominated by the delivery of essential routine physical care, with an absence of individualized care. The communication between the care partners, person with dementia, and healthcare team is often minimal, adding to the uncertainty about the plans for the future, which leads to frustration for both persons with dementia and their care partners. Whereas care partners are expected to be present at the bedside of the person with dementia to convey personal and medical information to the treating physician and healthcare team, this is not always possible as care partners are not always available nor do they know when the physicians do their rounds. As a result, the healthcare team may discuss important medical issues and medication management directly with the person with dementia (without the involvement of the care partners). The lack of information sharing with the care partners may put the person with dementia at risk, especially if the care plan involves high-risk investigations or medications. In this next example, a person with dementia is sent to the ED unaccompanied. The person is in the waiting room without assistance. The care partner appears distressed by the situation and lack of communication with the healthcare team.

> *Monday night, I get this phone call. She is upset, crying, confused um, she tells me that a cab took her to the hospital today and dropped her off and just left her. She doesn't know how she is gonna get back home. I don't know what's going on. She was totally confused. She was in the waiting room, and I said "How did you get a phone?" "Somebody let me use the hospital phone," she says. So she was able to ask, I guess after they triaged her, they just had her in the waiting room, in a wheel chair or something. She has been sitting there for hours. She was in excruciating pain and I don't know how she got somebody to get her to the phone.*

In addition, when only the person with dementia provides medical information, they may mistakenly provide inaccurate information because of the nature of their disease, which can impact their care. Given these challenges, it is important that care partners be actively engaged in the care planning for persons with dementia. However, multiple studies have reported that care partners feel they are not optimally included in the assessment, care planning, and information sharing [41–43]. Some care partners also reported that they had to repeat information multiple times, which was stressful and fatiguing [43].

As reported by this care partner, communication issues were striking in the emergency room:

> *They don't even read the chart about her dementia. They sent transfer papers with her but nobody reads it. Nobody reads it! It is emerge, they are busy, they don't have time to go through that thick binder that goes with her everywhere, but it should have a great big red or pink sticker on the front that said this woman has dementia, look in here.*

Another reason for communication issues is sensory impairments, which are often underestimated and underreported for persons with dementia [44]. Impaired vision and hearing influence a person with dementia's ability to engage and function in the acute care environment. Often, hearing aides are not brought to the hospital which impedes the person's ability to follow the interactions. Nurses, as well as hearing and vision experts, have expressed the need for education and suitable screening tools and formalized accountability within the screening process for vision and hearing loss in persons with dementia [45,46]. A growing body of research supports the need for appropriate screening tools to detect a sensory decline in this vulnerable population and the need to use sensory aides while in acute care facilities.

Finally, in persons with dementia, communication issues may result for people whose first language is not the local language and for whom cognitive decline can cause regression to the primary language and loss of language abilities [47]. As a result, hospitalized older adults with dementia may be unable to understand explanations, follow directions, or ask for help, which may have profound implications for patient outcomes [48]. Evidence exists in nursing homes that residents with dementia speaking English as a second language experience more agitation and neuropsychiatric symptoms than those speaking English as a first language [49]. With the number of persons with dementia with culturally and linguistically diverse backgrounds most likely to increase in the future, strategies for staff in hospitals working with this subsection of the population will need to be addressed, as it is not always possible to have a fluent speaker on hand.

In addition, the culture of efficiency and safety in hospitals is sometimes at odds with the care needs of persons with dementia [27]. In acute care, the focus is on minimizing risk, often to the detriment to the older adult with dementia. For example, leaving someone in bed, rather than having them walk and risk falling, can cause muscle strength loss and compromise an older person's ability to ambulate independently [50]. The risk for a fall has to be assessed in tandem with the risk from immobility [51]. Careful weighing of the advantages and disadvantages of keeping someone mobile with a possible risk of a fall must be made on a case-by-case basis.

Personal factors

Medical status

Globally, hospitalized persons with dementia may face increased risks of medical complications. Some complications may include adverse drug events, higher short-term mortality, urinary and/or fecal incontinence and poor pain management [52,53]. In addition, persons with dementia also have higher medical expenditures, length of stay, intensive care unit (ICU) stay, and in-hospital mortality [54]. Studies report that compared with medical patients without dementia, those with dementia had higher rates of urinary tract infection, pressure ulcer, pneumonia, deep vein thrombosis, sepsis, delirium, and failure to rescue (following sepsis, shock, gastrointestinal bleeding, deep vein thrombosis, or pneumonia) [55]. Surgical patients with dementia were reported to have significantly higher risks of acute renal failure, pneumonia, septicemia, stroke, and urinary tract infection [54]. These are all potentially preventable complications. An audit in Irish acute hospitals found that antipsychotic drugs were commonly prescribed to persons with dementia during hospitalization [56]. This is concerning, as antipsychotics have been associated with higher risk of adverse drug reactions in persons with dementia [57], the most severe of which are increased overall risk of cerebrovascular events and death [58].

Hospitalization is frequently associated with conditions related to pain, which brings challenges associated with its assessment and management, especially when caring for persons with dementia. An Irish national audit found that 75% of hospitalized patients with dementia received inadequate pain assessments [8]. People with dementia may have an atypical presentation of pain, which requires adapted assessment skills on the part of the healthcare professional and involvement of the care partner to ensure optimal pain management. Lack of recognition and treatment of pain may lead to agitation and a reduction in mobility resulting in functional decline. A qualitative study conducted in four acute care hospitals in England and Scotland investigated how pain is recognized, assessed, and managed in persons with dementia [59]. The investigators identified the following factors hindering pain assessment: communication barriers; organizational context; failure to build trust and rapport with patients; and failure to communicate with family regarding the patient's needs. Poor pain control adding onto the hospital's disorienting and distressing environment contributes to aggression and anxiety in persons with dementia, which in turn increase the risks of further cognitive decline and mortality [60,61].

Cognitive status

Persons with dementia are at a great risk of cognitive decline once admitted to an acute care hospital. Some acquire a superimposed delirium [28], as dementia is a strong predisposing factor for delirium. About 8% of persons with dementia will experience delirium during their course of hospitalization [62]. The risk for delirium increases to 50% in medical-surgical hospitalized patients with preexisting dementia [63]. Older adults with dementia are three to five times more likely to develop delirium than younger persons with dementia [60]. Development of delirium in the hospital setting is associated with poor long-term outcomes and in-hospital mortality in people who have preexisting physical and cognitive impairment [64]. A recent Italian study on the relationship between delirium, dementia, and in-hospital mortality suggested that the severity of cognitive impairment is associated with the risk of mortality [64]. Delirium is underrecognized, and diagnoses are missed in hospitalized patients, especially those with preexisting conditions. Studies report that up to 75% of nurses failed to identify delirium [65,66] and physicians' rate of misdiagnosis of delirium is 64.5% [67].

People with cognitive impairment (delirium or dementia) can experience aggression and agitation, which can be referred to as responsive behaviors [68] during their hospital stay. Responsive behaviors is a term, preferred by persons with dementia, representing how their behaviors, actions, gestures, and words have meaning and communicate needs or concerns. The meaning of their behaviors may be related to internal experiences, such as pain, or it may be related to the social or physical environment [69]. The majority of hospital clinicians and nurses express their low confidence and lack of knowledge to manage responsive behaviors in persons with dementia [70]. Unfortunately, patients with responsive behaviors are often chemically sedated, and thus their expressed needs are often not addressed. If the responsive behaviors are not properly managed, it will lead to further functional decline and increased length of hospital stay. Understanding the unmet needs of older adults with responsive behaviors and attending to their needs is required to improve the quality of care provided.

Functional status

Acute hospitalization often leads to functional decline, especially for persons with preexisting dementia [71,72]. During their hospital stay, persons with dementia are not always encouraged to be independent with their self-care. More specifically, the functional status of persons with dementia during hospitalization has been found to be significantly lower than it was 2 weeks before admission [71]. As a consequence, persons with dementia who were able to do their personal care before admission may become dependent on others for basic activities. The Alzheimer's Society of the United Kingdom highlighted the detrimental effects of hospital stay on the ability of persons with dementia to maintain their independence with their activities of daily living [73]. Unfortunately many will assume this is related to the trajectory of dementia when, in fact, it may be related to excess disability. This type of disability refers to the loss of an ability that comes from something other than the disease, such as inappropriate assessments, care or treatments.

There are common risk factors that contribute to the functional decline during hospitalization for persons with dementia [74]. For example, functional decline can be precipitated by environmental influences such as social isolation, or from iatrogenic effects of hospitalization such as side effects from medication, urinary catheters, intravenous, nasogastric tubes and dressings, isolation rooms due to infectious concerns, decreased nutritional intake and hydration, and decreased mobility. More specifically, decreased mobility can be caused by prolonged stay in bed because of various types of restraints [75]. Restraints can be physical such as having all bed rails up, the use of a tray table, mittens, or tight clothing. They can also be chemical, through medications such as antipsychotics and sedatives.

Care partners

When the person with dementia is hospitalized, care partners have many fears and concerns. Although the care partners expect that the person with dementia will be kept safe in the hospital, the carer often voices concerns about entrusting the well-being of their relative to others [15]. They tend to hold vigil to support their relatives' comfort and prevent complications while hospitalized.

As a daughter explained:

> *I left a little after midnight, 12:30 in the morning, when I had been there with her since 5 o'clock the previous morning and she had a nasogastric tube in her nose ... apparently when I left, she pulled that out 3 more times and her IV she pulled out a couple times. I know distinctly that when I left, I told them to give her something for pain, give her something to help her sleep so that she won't be bothered by the tubes, but the nurse was so busy and they didn't have anybody to kind of sit and watch with her. When I can back she was black and blue from starting IVs and everything. And I distinctly told them to call me if they needed me and they didn't. Because when I was sitting there, when someone was sitting with her she didn't have that behavior to pull things out. And it's very noisy, I'm sure it's very confusing.*

Routine evaluation of the care partner strain is essential when their relative with dementia is hospitalized, to inform the planning following their hospital stay [43]. When a person with dementia is hospitalized, additional strain exists for the caregiver as frequent visits to the hospital put demands on their time and interfere with their routines and responsibilities [43]. The care partners are often concerned about their ability to care for the person with dementia at hospital discharge. To mitigate their worry, care partners need to be heard and informed of the condition, diagnostics, and treatment of the person with dementia [43]. Researchers have found that at the time of discharge, care partners want be included in the planning process and to receive appropriate care instructions to reduce complications after discharge [15,76]. Care partners who were more affirmative in advocating for better discharge plans found it easier to access services and prevent inappropriate and unsafe care plans. Often, children and those with a healthcare background were able to advocate more successfully than a spouse [15].

Strategies for optimizing care of persons with dementia in hospitals

Over the last few decades, models of care have been created to improve hospital care for older adults. Despite the increased prevalence of persons with dementia being admitted into hospitals, most of the strategies that have been developed to optimize care in hospitals are focused on older adults and not specifically for older adults with dementia. However, the strategies described in this section can be adapted and also be helpful in caring for persons with dementia. In particular, a person-centered approach with someone with dementia is critical when utilizing the strategies listed in the following and will be highlighted in the final section.

Different strategies that have been developed to optimize care for older adults during the acute care stay include Comprehensive Geriatric Assessments, Senior-Friendly Hospitals/Care, Acute Care of the Elderly Units, 48/5 Pathway, Move On, Behavioral Support Teams, Family-Centered Function Focused Care, and a Rehabilitation Care Model for Older Adults with Cognitive Impairment.

Comprehensive geriatric assessments

Optimal disease management is the cornerstone of good care for persons with dementia. A comprehensive geriatric assessment, defined as a "multidimensional, interdisciplinary diagnostic process focused on determining an older adult's medical, psychological and functional capability to develop a coordinated and integrated plan for treatment and long-term follow-up" [77] has been shown to reduce short-term mortality, increase the chances of living at home at 1 year after discharge, and improve physical and cognitive function [78]. Effective management should start with a comprehensive assessment, followed by plan of care tailored to the needs of the individual. Multidimensional assessment carried out with the interdisciplinary team will assist in identifying the type of dementia, the stage, its impact on the person and their family, and the presence of other comorbidities. It will assist in guiding management of care for the individual. Each type of dementia has a characteristic cognitive and behavioral profile that will influence the nature of functional deficits [79], hence leading the team to target individual interventions within the care plan. Considerations of the comorbid status will help to optimize intrinsic capacity and function. The presence of other contributors such as depression may amplify the person with dementia's impairments, so thorough assessment and treatment is vital.

During these assessments, when the person has dementia, often the care partners will have more in-depth information regarding the patient's current and baseline functional status. By involving the care partner when making the assessment, they are less likely to become frustrated, stressed, and fatigued from attempting to advocate for themselves [15]. Their early recognition of symptoms of their relatives is critical in initiating medical care, especially because persons with dementia are often unable to recognize signs of their illness and need for attention [78]. Those without care partners are at more risk and vulnerable to poorer outcomes.

This thorough assessment should also lead to an understanding on the person's ability to hear and see and communicate with the healthcare team. For some, a pocket talker is a viable alternative to the more common hearing aids. Nurses commented that they had used the device with persons with dementia and it really helped with communication. Patients appeared less isolated and were more involved in the conversation [46]. Having more permissive visiting hours in acute care hospital has lifted some barriers to family presence, which can be useful if translation is required. If someone does not speak English, ensuring there are interpreters available is essential, especially if their family members are not available. Considering whether staff speak their language and planning shifts accordingly [49] and increasing the number of multicultural staff might mitigate many of the problems both providers and patients face during cross-cultural interactions [80]. Working with family members to provide staff with a few written words of basic vocabulary or using signs may help with interactions and care, as many persons with dementia can understand words and one-step instructions.

Senior-friendly hospitals/care

Senior-Friendly Hospital (SFH) is a framework for care which comprises five interrelated domains (organizational support, processes of care, emotional and behavioral environment, ethics in clinical care and research, and physical environment), with the aim of supporting hospitals to become senior-friendly [81]. This framework provided the blueprint for a self-assessment of all 155 adult hospitals across the province of Ontario in Canada. The system-wide analysis identified practice gaps and promising practices within each domain of the SFH framework. These results informed five domains to support hospitals at all stages of development in becoming friendly to older adults:

1. Organizational support, where there is leadership and support to make senior-friendly care a priority;
2. Process of care, focused on the provision of best evidence practices for older adult and delivering care in a manner that encourages continuity;
3. Emotional and behavioral environment, in that care is provided in a hospital that is free of ageism and respects the unique needs of individuals and their care partners;
4. Ethics in clinical care and research, where resources and capacity are available to address ethical issues that arise; and
5. Physical environment, where the structures, spaces, equipment, and facilities provide an environment that minimizes the vulnerabilities of frail individuals, promoting safety, independence, and functional well-being.

These priorities were identified, encouraging hospitals to implement or further develop their processes to better address hospital-acquired delirium and functional decline. These recommendations led to collaborative action across the province, including the development of an online toolkit and the identification of accountability indicators to support hospitals in quality improvement focusing on senior-friendly care.

Acute care of the elderly unit

The acute care of the elderly (ACE) model includes persons with dementia and has five components: (1) patient-centered care; (2) frequent medical review; (3) a prepared environment for enhancing mobility; (4) early rehabilitation; and (5) early discharge planning to prevent functional decline while in the hospital [82]. Implementation of the ACE components depends on the availability of healthcare professionals with adequate geriatric training and on collaboration among interdisciplinary team members to coordinate patient care. Results from previous studies demonstrated improved patient and system outcomes. Multidisciplinary services provided to older adults in the ACE model showed a significant decrease in mortality at three and 6 months after discharge [83]. Provision of ACE-based care contributes to improved functional outcomes in older adults during hospitalization and at discharge [84–86]. Some studies have also found significant reduction in the cost [87,88], hospital length of stay [89,90], and 30-day readmission for ACE patients after hospital discharge [89,91].

48/5 pathway

McElhaney et al. [92] created a 48/5 model of care that has preliminary evidence of its effectiveness for older adults with dementia. The 48/5 model is designed to minimize functional decline in older adults during acute hospitalization and return them to their baseline function. The 48/5 intervention addresses five key areas (delirium, medications, mobility, nutrition, and elimination) within 48 hours of hospital admission, and staff develop individualized care plans to improve hospital outcomes for older adults based on their assessments. The protocol focuses on

1. Nurses assessing the patient for delirium using the Confusion Assessment Method and treating the underlying causes;
2. Nurses with pharmacists reviewing medication and contacting MDs if there are nonrecommended medications;
3. Encouraging patients to move and walk to the dining room for meals;
4. Monitoring nutritional status and hydration and providing supplements if required; and
5. Monitoring bowel and bladder regulation, providing laxatives if no bowel movement for three consecutive days, and removing indwelling catheters if not medically necessary.

The 48/5 approach is triggered through a geriatric assessment screening tool and has been implemented using a quality improvement strategy to support practice change within care teams. The 48/5 implementation pathway consists of four components: (1) screening assessments; (2) daily huddles to report on the assessments findings; (3) deliver interventions once needs are assessed; and (4) perform daily reassessments to determine if the interventions are effective. The staff members are supported through peer interactions and by the project coordinator.

Move on

A key intervention to prevent functional decline and optimize patient outcomes is the promotion of early mobilization for PWD admitted to acute care settings [93]. Early mobilization will prevent dependency on others and the risk of deconditioning that is commonly seen when a person is immobile for a long period. Early mobilization is, in general, used for all patients, but it can be argued that it needs to be promoted even more for the population of persons with dementia because they are at a higher risk of functional decline when hospitalized. The Move On intervention consists of early mobilization promotion, defined as assessing patient for mobility and functional status within 24 h of admission and encouraging appropriate activities immediately. Evaluation of the Move On intervention is ongoing.

Behavioral support teams

Behavioral support teams have been introduced in Ontario, Canada, healthcare settings. The focus of the team is on supporting the care and transition of patients who are currently admitted to acute care hospitals and experiencing responsive behaviors that are creating barriers to the provision of care and transition from the hospital. This includes patients with responsive behaviors related to cognitive impairment, mental illness, and/or substance use [94]. These nurses develop an evidence-based behavioral support plan to support other team members in caring for these individuals and their families and assist in successful transitions to other living destinations. As this is a recent initiative, no formal evaluations have been completed in acute care hospitals. Based on informal reports, there appears to be merit in this initiative.

Family-centered function focused care

Boltz et al. [74] recognized that family care partners can exert considerable influence over the delivery of care to persons with dementia when hospitalized, as older adults are more likely to engage in self-care and walking programs when families encourage them to do so. Therefore to capitalize on this reality for those who have family care partners and to promote the best possible postdischarge outcomes, Boltz et al. created a Family-Centered Function Focused Care (Fam-FFC) program to promote functional recovery in hospitalized older adults with dementia. The Fam-FFC program includes (1) staff education on care of older adults with cognitive challenges; (2) family/patient education; (3) jointly developed bedside goals and treatment plans; and (4) postacute follow-up including a home visit within 48 hours of discharge. The model appears promising. Initial testing demonstrated less severity and shorter duration of delirium, better ADL, and better walking performance [74]. In addition, there was increased preparedness for caregiving and a decrease in their anxiety and depression. Care partners presence is reassuring for the person with dementia who has difficulty adjusting to the new, busy, and noisy environment of the hospital. This development of an innovative model of care that focusses on the carer and persons with dementia is a positive step forward.

In terms of patient and family education and knowing what to expect when hospitalized, the National Institute on Aging has created a website for care partners to access, which is a very good resource if a relative or friend with dementia has to be hospitalized [95]. Suggestions for the care partner such as what to expect during the hospitalization and tips for working with the hospital staff are provided to help make it a smoother transition. As a trip to the hospital can be stressful for the person with dementia and the care partner, being prepared may make the visit to the emergency room or hospital easier. In addition, keeping in mind that hospitals are not designed for persons with dementia, care partners are encouraged to discuss with their healthcare professionals if avoiding a trip to the hospital is at all possible.

Rehabilitation care model for older adults with cognitive impairment

Despite a widespread nihilistic belief that persons with dementia cannot be successfully rehabilitated, resulting in limited access to rehabilitation, McGilton et al. [40] created a patient-centered rehabilitation model of care for individuals with cognitive impairment (PCRM-CI). This model combines intensive rehabilitation with delirium and dementia management, and concurrent education and support for healthcare providers and family members. The evaluation of the program revealed that individuals with cognitive impairment (including those with dementia) in the intervention group were more likely to be discharged home than those in the control group after a hip fracture [96]. Six months later, older adults were more likely to ambulate inside and outside, and go shopping. Of note, preadmission functional impairment was more strongly associated with poor outcomes than cognitive impairment [97]. In fact, patients with hip fracture and a diagnosis of dementia can respond well to more intensive rehabilitation settings, showing better outcomes (living arrangements, reduced length of stay, and functional gain) than achieved with less intense inpatient rehabilitation programs or no rehabilitation [98]. Furthermore, there is no evidence that involvement in rehabilitation results in harm to participants, nor that individuals with cognitive impairment are unable to participate [99,100]. This nihilistic belief that a diagnosis of dementia makes the person unable to participate effectively and benefit from a rehabilitation program can lead to reluctance on the part of staff and administrators to devote scarce resources to patients who are cognitively impaired, no matter where they reside [101]. Clearly, this perception must change, and rehabilitation must start in acute care. This may be one way to prevent delayed discharges.

Approaches to care of older adults with dementia

While models of care grounded in good gerontological principles are essential for persons with dementia, approaches that staff use when working with this population are also essential. Person-centered care (PCC) has been identified as the ideal approach to caring for persons with dementia [36]. A PCC approach places priority on the relationship with the individual rather than on the completion of tasks. However, many person-centered approaches or models are not clearly operationalized. An exemption is a model developed by McGilton et al. [102], a Model for Excellence in Dementia Care, which has been reframed over time as the REAP model (Relating Well, Environment Manipulation, Abilities Focus, and Personhood). The manner in which healthcare professionals relate to persons with dementia in hospital is important during the provision of individualized care. Specific relational behaviors of staff have been linked to positive outcomes for persons with dementia [103]. Examples of these relational approaches include, but are not limited to, maintaining close proximity, utilizing various forms of touch that are comfortable for the patient, hesitating in care when necessary, being flexible, and acknowledging the client's subjective experiences. Manipulation of the environment to meet the unique needs of a person with dementia is an effective strategy when personalizing care. The environment plays an important role in reducing responsive behaviors and improving quality of life of persons with dementia [104]. Environmental strategies may optimize an individual's performance of activities of daily living (ADLs) such as placing frequently used items in a specific location or improving the person's ability to eat with increased light [105]. There is evidence to suggest that the abilities are retained to a moderate or high level in persons with dementia, so healthcare professionals must promote those abilities that are found to be preserved and compensate for those that are not [106]. The ability to recognize retained abilities, vs. disabilities, may help to prevent or reverse excess disability, thus promoting function [107]. Finally, knowing the person involves becoming familiar with the individual and gaining an understanding of who they are as individuals and using this information when providing care. By knowing the individual, one is more likely to understand why individuals have responsive behaviors. As Dupuis et al. [68] state, instead of judging behaviors of persons with dementia, the emphasis should be on viewing their behaviors as actions and meaningful responses with an attempt to connect and understand their meaning. There are three guiding principles for understanding responsive behaviors: "(1) all personal expressions (words, gestures, actions) have meaning; (2) personal expressions communicate meanings, needs and concerns; and (3) to understand their meaning, you must consider the factors influencing his behavior (physical, emotional and environmental elements etc.)" [69]. Staff are called to be truly present, actively listen and understand the meaning in their actions and responses, and attempt to meet any unmet needs (such as being cold, in pain, lonely, or constipated). Family care partners and friends are also important sources for getting to know the individual and their responses.

Preventing hospitalizations

Many of the care approaches described in this chapter could be implemented to improve the hospital experience of persons with dementia and their care partners. However, policies to support and fund implementation of these practices are needed from healthcare provider organizations and funders. Similarly, policies could be implemented to prevent unnecessary hospitalization. Two programs will be highlighted that were developed with this aim in mind.

The *Geriatric Emergency Management Nurse (GEM Nurse)* program was developed several years ago to locate GEM nurses into emergency departments, with goals to [1] improve delivery of care through the development of unique, site-appropriate solutions, and [2] prevent unnecessary hospital admissions. GEM nurses incorporate capacity building into their role, to develop and strengthen the skills, abilities of staff and the processes, and resources of emergency departments [108]. Care processes focus on areas of staffing, mobilization, comfort, medication, hygiene, nutrition/hydration, cognition, environment, equipment, and stimulation. GEM nurses work collaboratively with the emergency department team to facilitate change in the way that emergency department care is provided to older persons experiencing health emergencies. These nurses promote comprehensive holistic care, redirect unnecessary admissions or recidivism, and provide direct communication between families, patients, and other agencies [109]. The GEM nurse plays a crucial role between linking the senior to community services and healthcare outcomes. In a 1-year study,

GEM nurses in eight Ontario emergency departments were involved in nearly 3000 visits by high-risk seniors [110]. Early reports of the GEM nurse model offered supportive data, intriguing patient accounts and positive clinical outcomes.

Most recently, emergency medical services (EMS) are being explored in the role they could play in unplanned, urgent, and emergency care for older people with dementia [111]. There is an increasing international interest in the role of EMS to avoid transportation of older adults with dementia to emergency if deemed not medically necessary. A protocol was developed investigating the reasons for emergency calls including those from persons with dementia or care partners, with the aim of developing an assessment to reduce hospital admissions [112]. There will be a need for interactions between EMS personnel and community services to fulfill this mandate, and future research will focus on how to make this happen.

Finally, the message for policymakers, practitioners, families, and persons with dementia needs to be "living well with dementia," with a focus on maintaining function for as long as possible, regaining lost function when there is the potential to do so, and adapting to lost function that cannot be regained in all settings including acute care. Service delivery of persons with dementia must be reoriented such that evidence-based reablement approaches are integrated into routine care across all sectors [81]. To achieve this vision, education of healthcare professionals needs to be updated to meet the needs of persons with dementia so that they receive appropriate care while hospitalized.

Building research evidence

Utilization of evidence-based recommendations in care of persons with dementia is essential to promote the best outcomes. Areas of future research should include studies that are generalizable to the practice setting, as well as research focusing on the translation of research into practice for healthcare professionals, the development of strategies for effective implementation of evidence-based practice recommendations, and examination of the role of management in sustained implementation of new knowledge into practice.

Conclusions

Dementia is common among hospitalized older adults. People with dementia are at an increased risk of adverse outcomes and of functional decline once hospitalized. However, decline is not inevitable. Inclusion of care partners in care planning and provision is essential. Programs are surfacing on ways to decrease acute hospitalization risks, although widespread adoption has not yet been achieved.

To all the individuals affected by dementia, may this work help your voice be heard and may your experiences come to light in making this journey through acute care hospital better.

References

[1] Phelan A, Borson S, Grothaus L, Balch S, Larson EB. Association of incident dementia with hospitalizations. J Am Med Assoc 2012;307(2):165−72.
[2] Bynum JP, Rabins PV, Weller W, Niefeld M, Anderson GF, Wu AW. The relationship between a dementia diagnosis, chronic illness, medicare expenditures, and hospital use. J Am Geriatr Soc 2004;52(2):187−94.
[3] Gill SS, Camacho X, Poss JW. Community dwelling adults with dementia: tracking encounters with the health system. In: Bronskill SE, Camacho X, Gruneir A, Ho MM, editors. Health system use by frail Ontario seniors: an in-depth examination of four vulnerable cohorts. Toronto, ON: Institute for Clinical Evaluative Sciences; 2011. p. 47−68.
[4] Canadian Association for Retired Persons (Internet). Wait times alliance. Shedding light on Canadians'. Report Card on Wait Times in Canada. 2012. Available from: www.carp.ca/2012/07/27/shedding-light-on-canadians-total-wait-for-care-wait-time-alliance/.
[5] Mukadam N, Sampson EL. A systematic review of the prevalence, associations and outcomes of dementia in older general hospital inpatients. Int Psychogeriatr 2011;23(3):344−55.
[6] Department of Health (Internet). National service framework: older people. London: The Stationary Office; 2001. Available from: www.gov.uk/government/publications/quality-standards-for-care-services-for-older-people.
[7] Silverstein M, Maslow K. Improving hospital care for people with dementia. New York: Springer Publishing Co; 2006.
[8] Timmons S, O'Shea E, O'Neill D, Gallagher P, de Siún A, McArdle D, Gibbons P, Kennelly S. Acute hospital dementia care: results from a national audit. BMC Geriatr 2016;16:113. https://doi.org/10.1186/s12877-016-0293-3.
[9] Hickey A, Clinch D, Groarke EP. Prevalence of cognitive impairment in the hospitalised elderly. Int J Geriatr Psychiatry 1997;12(1):27−33.
[10] Jacques A, Jackson G. Understanding dementia. 3rd ed. Edinburgh: Churchill Livingstone; 1999.
[11] Moyle W, Borbasi S, Wallis M, Olorenshaw R, Gracia N. Acute care management of older people with dementia: a qualitative perspective. J Clin Nurs 2011;20(34):420−8.
[12] Goodall D. Environmental changes increase hospital safety for dementia patients. Holist Nurs Pract 2006;20(2):80−4.
[13] Draper B, Karmel R, Gibson D, Peut A, Anderson P. The hospital dementia services project: age differences in hospital stays for older people with and without dementia. Int Psychogeriatr 2011;23(10):1649−58.
[14] Alzheimer Society of Canada. Rising tide: the impact of dementia on Canadian society, Toronto, Ontario. 2010. Available from: alzheimer.ca/sites/default/files/files/national/advocacy/asc_rising_tide_full_report_e.pdf.

[15] (Internet) Lemay G. The experience of people with dementia and their caregivers during acute hospitalization (dissertation). Ottawa (ON): University of Ottawa; 2014. Available from: ruor.uottawa.ca/bitstream/10393/31451/1/Lemay_Genevieve_2014_thesis.pdf.
[16] Nourhashemi F, Andrieu S, Sastres N, Ducassé D, Lauque A, Sinclair, et al. Descriptive analysis of emergency hospital admissions of patients with Alzheimer's disease. Alzheimer Dis Assoc Disord 2001;15(1):21–5.
[17] Rao A, Suliman A, Vuik S, Aylin P, Darzi A. Outcomes of dementia: systematic review and meta-analysis of hospital administrative database studies. Arch Gerontol Geriatr 2016;66:198–204. https://doi.org/10.1016/j.archger.2016.06.008.
[18] Morrison RS, Siu AL. Survival in end-stage dementia following acute illness. JAMA 2000;284(1):47–52. https://doi.org/10.1001/jama.284.1.47.
[19] Sampson E, Blanchard M, Jones L, Tookman A, King M. Dementia in the acute hospital: prospective cohort study of prevalence and mortality. Br J Psychiatry 2009;195(1):61–6.
[20] Andrieu S, Reynish M, Nourhashemi F, Shakespeare A, Moulias S, Ousset PJ, et al. Predictive factors of acute hospitalizations in 134 patients with Alzheimer's disease: a prospective one year study. Int J Geriatr Psychiatry 2002;17(5):422–6.
[21] Voisin T, Andrieu S, Cantet C, Vellas B, REAL.FR G. Predictive factors of hospitalizations in Alzheimer's disease: a two-year prospective study in 686 patients of the REAL.FR study. J Nutr Health Aging 2010;14(4):288–91.
[22] Toot S, Devine M, Akporobaro A, Orrell M. Causes of hospital admission for people with dementia: a systematic review and meta-analysis. J Am Med Assoc 2013;14(7):463–70.
[23] Tuppin P, Kusnik-Joinville O, Weill A, Ricordeau P, Allemand H. Primary health care use and reasons for hospital admissions in dementia patients in France: database study in 2007. Dement Geriatr Cognit Disord 2009;28(3):225–32.
[24] Inouye SK, Bogardus Jr ST, Charpentier PA, Leo-Summers L, Acampora D, Holford TR, et al. A multicomponent intervention to prevent delirium in hospitalized older patients. N Engl J Med 1999;340(9):669–76.
[25] Witlox J, Eurelings LM, de Jonghe JM, Kalisvaart KJ, Eikelenboom P, van Gool WA. Delirium in elderly patients and the risk of postdischarge mortality, institutionalization, and dementia: a meta-analysis. JAMA 2010;304(4):443–51.
[26] Miller EA, Rosenheck RA, Schneider LS. Caregiver burden, health utilities, and institutional service use in Alzheimer's disease. Int J Geriatr Psychiatry 2012;27(4):382–93.
[27] Parke B, Hunter KF, Schulz ME, Jouanne L. Know me – a new person-centered approach for dementia-friendly emergency department care. Dementia (London) 2016;18(2):432–47. https://doi.org/10.1177/1471301216675670.
[28] Devere R. Short-and long-term cognitive consequences of acute non-neurological hospitalization. Pract Neurol January 2012:30–5.
[29] Bauer M, Fitzgerald L, Koch S. Hospital discharge as experienced by family carers of people with dementia: a case for quality improvement. J Healthc Qual 2011;33(6):9–16.
[30] Fitzgerald LR, Bauer M, Koch M, King SJ. Hospital discharge: recommendations for performance improvement for family carers of people with dementia. Aust Health Rev 2011;35(3):364–70.
[31] Naylor MD, Hirschman KB, Bowles KH, Bixby MB, Konick-McMahan J, Stephens C. Care coordination for cognitively impaired older adults and their caregivers. Home Health Care Serv Q 2007;26(4):57–78.

[32] Costa AP, Poss JW, Peirce T, Hirdes JP. Acute care inpatients with long-term delayed-discharge: evidence from a Canadian health region. BMC Health Serv Res 2012;12(1):172.
[33] Kuluski K, Im J, McGeown M. "It's a waiting game" a qualitative study of the experience of carers of patients who require an alternate level of care. BMC Health Serv Res 2017;17(1):318.
[34] Canadian Institute for Health Information (Internet). Definitions and guidelines to support ALC designation in acute inpatient care introduction guidelines to support ALC designation by clinicians. Canadian Institute for Health Information; 2016. p. 1–3. Available from: www.cihi.ca/sites/default/files/document/acuteinpatientalc-definitionsandguidelines_en.pdf.
[35] (Internet) Pringle D. Aging and the health care system: am I in the right queue? National Advisory Council on Aging. Ottawa: Minister of Public Works and Government Services Canada; 1998. Available from: www.publications.gc.ca/site/eng/9.559136/publication.html.
[36] Clissett P, Porock D, Harwood RH, Gladman JR. The challenges of achieving person-centred care in acute hospitals: a qualitative study of people with dementia and their families. Int J Nurs Stud 2013;50(11):1495–503.
[37] McBrien B. Translating change: the development of a person-centred triage training programme for emergency nurses. Int Emerg Nur. 2009;17(1):31–7.
[38] Institute of Medicine. Retooling for an aging America: building the health care workforce. Washington (DC): National Academies Press; 2008.
[39] Teodorczuk A, Mukaetova-Ladinska E, Corbett S, Welfare M. Deconstructing dementia and delirium hospital practice: using cultural historical activity theory to inform education approaches. Adv Health Sci Educ Theory Pract 2015;20(3):745–64. https://doi.org/10.1007/s10459-014-9562-0.
[40] McGilton K, Davis D, Mahomed N, Flannery J, Jaglal S, Cott C, Naglie G, et al. An inpatient rehabilitation model of care targeting patients with cognitive impairment. BMC Geriatr 2012;12(21).
[41] Bloomer M, Digby R, Tan H, Crawford K, Williams A. The experience of family carers of people with dementia who are hospitalised. Dementia (London) 2016;15(5):1234–45. https://doi.org/10.1177/1471301214558308.
[42] Hynninen N, Saarnio R, Isola A. Treatment of older people with dementia in surgical wards from the viewpoints of the patients and close relatives. J Clin Nurs 2015;24(23–24):3691–9. https://doi.org/10.1111/jocn.13004.
[43] Boltz M, Chippendale T, Resnick B, Galvin JE. Anxiety in family caregivers of hospitalized persons with dementia: contributing factors and responses. Alzheimer Dis Assoc Disord 2015;29(3):236–41.
[44] O'Malley PG. Evolving insights about the impact of sensory deficits in the elderly. JAMA Intern Med 2013;173(4):299.
[45] Hobler F, Argueta-Warden X, Rodríguez-Monforte M, Escrig-Pinol A, Wittich W, McGilton KS. Exploring the sensory screening experiences of nurses working in long-term care homes with residents who have dementia: a qualitative study. BMC Geriatr 2018;18(1):235. https://doi.org/10.1186/s12877-018-0917-x.
[46] Wittich W, Hobler F, Jarry J, McGilton KS. Recommendations for successful sensory screening in older adults with dementia in long-term care: a qualitative environmental scan of Canadian specialists. BMJ Open 2018;8(1):e019451. https://doi.org/10.1136/bmjopen-2017-019451.
[47] McMurtray A, Saito E, Nakamoto B. Language preference and development of dementia among bilingual individuals. Hawaii Med J 2009;68(9):223–6.

[48] Miller CA. Communication difficulties in hospitalized older adults with dementia. Am J Nurs 2008;108(3):58–66. https://doi.org/10.1097/01.NAJ.0000311828.13935.1e.
[49] Cooper C, Rapaport P, Robertson S, Marston L, Barber J, Manela M, et al. Relationship between speaking English as a second language and agitation in people with dementia living in care homes: results from the MARQUE (Managing Agitation and Raising Quality of life) English national care home survey. Int J Geriatr Psychiatry 2018;33(3):504–9. https://doi.org/10.1002/gps.4786.
[50] Kortebein P, Symons B, Ferrando A, Paddon-Jones D, Ronsen O, Protas E, et al. Functional impact of 10 days of bed rest in healthy older adults. J Gerontol 2008;63(10):1076–81. https://doi.org/10.1093/gerona/63.10.1076.
[51] Ahlskog JE, Geda YE, Graff-Radford NR, Petersen RC. Physical exercise as a preventive or disease modifying treatment of dementia and brain aging. Mayo Clin Proc 2011;86(9):876–84.
[52] Pi HY, Gao Y, Wang J, Hu MM, Nie D, Peng PP. Risk factors for in-hospital complications of fall-related fractures among older Chinese: a retrospective study. BioMed Res Int 2016;2016:8612143.
[53] Liao JN, Chao TF, Liu CJ, Wang KL, Chen SJ, Tuan TC. Risk and prediction of dementia in patients with atrial fibrillation—a nationwide population-based cohort study. Int J Cardiol 2015;199:25–30. https://doi.org/10.1016/j.ijcard.2015.06.170.
[54] Hu CJ, Liao CC, Chang CC, Wu CH, Chen TL. Postoperative adverse outcomes in surgical patients with dementia: a retrospective cohort study. World J Surg 2012;36(9):2051–8. https://doi.org/10.1007/s00268-012-1609x.
[55] Bail K, Berry H, Grealish L, Draper B, Karmel R, Gibson D, et al. Potentially preventable complications of urinary tract infections, pressure areas, pneumonia, and delirium in hospitalised dementia patients: retrospective cohort study. BMJ Open 2013;3(6):e002770. https://doi.org/10.1136/bmjopen-2013-002770.
[56] Gallagher P, Curtin D, de Siún A, O'Shea E, Kennelly S, O'Neill D, et al. Antipsychotic prescription amongst hospitalized patients with dementia. QJM 2016;109(9):589–93. https://doi.org/10.1093/qjmed/hcw023.
[57] Kanagaratnam L, Dramé M, Trenque T, Oubaya N, Nazeyrollas P, Novella JL, et al. Adverse drug reactions in elderly patients with cognitive disorders: a systematic review. Maturitas 2016;85:56–63. https://doi.org/10.1016/j.maturitas.2015.12.013.
[58] Schneider LS, Dagerman K, Insel PS. Efficacy and adverse effects of atypical antipsychotics for dementia: meta-analysis of randomized, placebo-controlled trials. Am J Geriatr Psychiatry 2006;14(3):191–210.
[59] Lichtner V, Dowding D, Allcock N, Keady J, Sampson EL, Briggs M, et al. The assessment and management of pain in patients with dementia in hospital settings: a multi-case exploratory study from a decision making perspective. BMC Health Serv Res 2016;16(1):427. https://doi.org/10.1186/s12913-016-1690-1.
[60] Sampson EL, White N, Lord K, Leurent B, Vickerstaff V, Scott S, et al. Pain, agitation, and behavioural problems in people with dementia admitted to general hospital wards: a longitudinal cohort study. Pain 2015;156(4):675–83.
[61] Cunningham C. Managing pain in patients with dementia in hospital. Nurs Stand 2006;20(46):54–8. https://doi.org/10.7748/ns2006.07.20.46.54.c4473.
[62] Travers C, Byrne G, Pachana N, Klein K, Gray L. Prospective observational study of dementia and delirium in the acute hospital setting. J Intern Med 2013;43(3):262–9.
[63] Robinson TN, Raeburn CD, Tran ZV, Angles EM, Brenner LA, Moss M. Postoperative delirium in elderly: risk factors and outcomes. Ann Surg 2009;249(1):173–8.

[64] Morandi A, Di Santo S, Zambon A, Mazzone A, Cherubini A, Mossello E, et al. Delirium, dementia and in-hospital mortality: the results from the Italian delirium day 2016, a national multicenter study. J Gerontol A Biol Sci Med Sci July 4, 2018. https://doi.org/10.1093/gerona/gly154 [Epub ahead of print)].
[65] Australian Institute of Health and Welfare (Internet). Dementia care in hospitals: costs and strategies. Canberra. 2013. Available from: https://www.aihw.gov.au/reports/aged-care/dementia-care-in-hospitals-costs-and-strategies/contents/table-of-contents.
[66] Rice KL, Gomez M, Theall KP, Knight M, Foreman MD. Nurses' recognition of delirium in the hospitalized older adult. Clin Nurse Spec 2011;25(6). 299-11.
[67] Ritter S, Cardoso A, Lins M, Zoccoli TL, Freitas MP, Camargos EF. Underdiagnosis of delirium in the elderly in acute care hospital settings: lessons not learned. Psychogeriatrics 2018;18(4):268−75. https://doi.org/10.1111/psyg.12324.
[68] Dupuis S, Wiersma E, Loiselle L. Pathologizing behavior: meanings of behaviors in dementia care. J Aging Stud 2012;26:162−73.
[69] Alzheimer Society of Ontario (Internet). Examples of responsive behaviours. 2017. Available from: alzheimer.ca/en/on/We-can-help/Resources/Shifting-Focus/Examples-of-responsive-behaviour.
[70] Temple A. CHOPS: care of the confused hospitalised older person study. Sydney: NSW Agency for Clinical Innovation; 2013.
[71] Covinsky K, Palmer R, Fortinsky R, Counsell S, Stewart A, Kresevic D, et al. Loss of independence in activities of daily living in older adults hospitalized with medical illnesses: increased vulnerability with age. J Am Geriatr Soc 2003;51(4):451−8.
[72] Hartley P, Alexander K, Adamson J, Cunningham C, Embleton G, Romero-Ortuno R. Association of cognition with functional trajectories in patients admitted to geriatric wards: a retrospective observational study. Geriatr Gerontol Int 2017;17(10):1438−43.
[73] Alzheimer's Society (Internet). Counting the cost: caring for people with dementia in hospital wards. 2009. Available from: www.alzheimers.org.uk/sites/default/files/2018-05/Counting_the_cost_report.pd.f.
[74] Boltz M, Chippendale T, Resnick B, Galvin JE. Testing family-centered, function-focused care in hospitalized persons with dementia. Neurodegener Dis Manag 2015;5(3):203−15.
[75] Nakanishi M, Okumura Y, Ogawa A. Physical restraint to patients with dementia in acute physical care settings: effect of the financial incentive to acute care hospitals. Int Psychogeriatr 2018;30(7):991−1000.
[76] (Internet) Donnelly M, McElhaney J, Carr M. Improving BC's care for people with dementia in the emergency department and acute hospital. 2011. Available from: www2.gov.bc.ca/assets/gov/topic/AE132538BBF7FAA2EF5129B860EFAA4E/pdf/improvingcaredementiareport2011.pdf.
[77] Rubenstein LZ, Stuck AE, Siu AL, Wieland D. Impacts of geriatric evaluation and management programs on defined outcomes: overview of the evidence. J Am Geriatr Soc 1991;39(9 Pt 2):8S−16S.
[78] Ellis G, Langhorne P. Comprehensive geriatric assessment for older hospital patients. Br Med Bull 2005;71(1):45−59. https://doi.org/10.1093/bmb/ldh033.
[79] Poulos CJ, Bayer A, Beaupre L, et al. A comprehensive approach to reablement in dementia. Alzheimers Dement (NY) 2017;3(3):450−8. https://doi.org/10.1016/j.trci.2017.06.005.

[80] Komaric N, Bedford S, van Driel ML. Two sides of the coin: patient and provider perceptions of health care delivery to patients from culturally and linguistically diverse backgrounds. BMC Health Serv Res 2012;12:322. https://doi.org/10.1186/1472-6963-12-322.
[81] Wong K, Ryan D, Liu B. A system-wide analysis using a senior-friendly hospital framework identifies current practices and opportunities for improvement in the care of hospitalized older adults. J Am Geriatr Soc 2014;62(11):2163–70. https://doi.org/10.1111/jgs.13097.
[82] Wong RY, Shaw M, Acton C, Wilbur K, McMillan M, Breurkens E, et al. An interdisciplinary approach to optimize health services in a specialized acute care for elders unit. Geriatr Today 2003;6(3):177–86.
[83] Salvedt I, Mo ES, Fayers P, Sletvold O. Reduced mortality in treating acutely sick frail older patients in geriatric evaluation and management unit: a prospective randomized trial. J Am Geriatr Soc 2002;50(5):792–8.
[84] Perez-Zepeda MU, Gutierez-Robledo LM, Sanchez-Garcia S, Juarez-Cedillo T, Gonzalez JJ, Franco-Marina F, et al. Comparison of a geriatric unit with a general ward in Mexican elders. Arch Gerontol Geriatr 2012;54:e370–5.
[85] Wald HL, Glasheen JJ, Guerrasio J, Youngwerth JM, Cumbler EU. Evaluation of a hospitalist-run acute care for the elderly service. J Hosp Med 2011;6(6):313–21.
[86] Zelada MA, Salinas R, Baztan JJ. Reduction of functional deterioration during hospitalization in an acute geriatric unit. Arch Gerontol Geriatr 2009;48(1):35–9.
[87] Barnes DE, Palmer RM, Kresevic DM, Fortinksy RH, Kowal J, Chren M. Acute care for elders units produced shorter hospital stays at lower cost while maintaining patients' functional status. Health Aff (Millwood) 2012;31(6):1227–36.
[88] Flood K, MacLennan P, McGrew D, Green D, Dodd C, Brown C. Effects of an acute care for elders unit on costs and 30-day readmissions. JAMA Int Med 2013;173(11):981–7. https://doi.org/10.1001/jamainternmed.2013.524.
[89] Allen KR, Hazelett SE, Palmer RR, Jarjoura DG, Wickstrom GC, Weinhardt JA, et al. Developing a stroke unit using the acute care for elders intervention and model of care. J Am Geriatr Soc 2003;51(11):1660–7.
[90] Ahmed N, Taylor K, McDaniel Y, Dyer CB. The role of an acute care for the elderly unit in achieving hospital quality indicators while caring for frail hospitalized elders. Popul Health Manag 2012;15(4):236–40.
[91] Jayadevappa R, Chhatre S, Weiner M, Raziano DB. Health resource utilization and medical care cost of acute care elderly unit patients. Value Health 2006;9(3):186–92.
[92] Canadian Foundation for Healthcare Improvement (CFHI) (Internet). Improving treatment for seniors in acute care. Toronto. 2014. Available from: www.cfhi-fcass.ca/OurImpact/ImpactStories/ImpactStory/2013/06/16/b1c07b79-915b-4c66-b8b6-91488a590a8b.aspxe.
[93] Liu B, Almaawiy U, Moore JE, Chan W, Straus SE. Evaluation of a multisite educational intervention to improve mobilization of older patients in hospital: protocol for mobilization of vulnerable elders in Ontario (MOVE ON). BMC Implement Sci 2013;8:76.
[94] Gutmanis I, Harvey D, Hillier L, LeClair K. Health care redesign for responsive behaviours—the behavioural supports Ontario experience: lessons learned and keys to success. Can J Community Ment Health 2015;34(1):45–63.
[95] National Institute on Aging (Internet). Going to the hospital: tips for dementia caregivers. Maryland. 2017. Available from: www.nia.nih.gov/health/going-hospital-tips-dementia-caregivers.
[96] McGilton K, Davis A, Naglie G, Mahomed N, Flannery J, Jaglal S, et al. Evaluation of patient-centered rehabilitation model targeting older persons with a hip fracture,

including those with cognitive impairment. BMC Geriatr 2013;13:136. https://doi.org/10.1186/1471-2318-13-136.
[97] McGilton K, Chu C, Naglie G, Wyk P, Stewart S. Factors influencing outcomes of older adults after undergoing rehabilitation for hip fracture. JACS 2016;64(8):1601–9.
[98] Seitz DP, Gill SS, Austin PC, Bell CM, Anderson GM, Gruneir A, et al. Rehabilitation of older adults with dementia after hip fracture. J Am Geriatr Soc 2016;64(1):47–54. https://doi.org/10.1111/jgs.13881.
[99] Allen J, Koziak A, Buddingh S, Liang J, Buckingham J, Beaupre LA. Rehabilitation in patients with dementia following hip fracture: a systematic review. Physiother Can 2012;64(2):190–201.
[100] Resnick B, Beaupre L, McGilton KS, Galik E, Liu W, Neuman MD, et al. Rehabilitation interventions for older individuals with cognitive impairment post-hip fracture: a systematic review. J Am Med Dir Assoc 2016;17(3):200–5.
[101] McFarlane RA, Isbel ST, Jamieson MI. Factors determining eligibility and access to subacute rehabilitation for elderly people with dementia and hip fracture. Dementia (London) 2017;16(4):413–23. https://doi.org/10.1177/1471301215599704.
[102] McGilton K, Lever J, Mowat J, Parnell I, Perivolaris A, Biscardi M. Guideline recommendations to improve dementia care. Alzheimers Care Q 2007;8(2):109–15.
[103] McGilton KS, Sidani S, Boscart VM, Guruge S, Brown M. The relationship between care providers' relational behaviors and residents mood and behavior in long-term care settings. Aging Ment Health 2012;16(4):507–15. https://doi.org/10.1080/13607863.2011.628980.
[104] Zeisel J, Raia P. Nonpharmacological treatment for Alzheimer's disease: a mind-brain approach. Am J Alzheimers Dis Other Demen 2000;15(6):331–40.
[105] Cohen-Mansfield J. Nonpharmacological interventions for persons with dementia. Alzheimers Care Q 2005;6(2):129–45.
[106] Wells D, Dawson P. Description of retained abilities in older persons with dementia. Res Nurs Health 2000;23(2):158–66.
[107] Wells D, Dawson P, Sidani S, Craig D, Pringle D. Effects of an abilities focused program of morning care on residents who have dementia and on caregivers. J Am Geriatr Soc 2000;48(4):442–9.
[108] Flynn D, Jennings J, Moghabghab R, Nancekivell T, Tsang C, Cleland M, et al. Raising the bar of care for older people in Ontario emergency departments. Int J Older People Nurs 2010;5(3):219–26. https://doi.org/10.1111/j.1748-3743.2010.00209.x.
[109] Di Sabatino S. Perspectives (Pre-2012). Toronto 2009;33(3):18–22.
[110] Regional Geriatric Program of Toronto (Internet). A collaborative model for service delivery in the Emergency Department. Toronto. 2009. Available from: https://www.rgptoronto.ca/wp-content/uploads/2017/12/Collaborative_Model_of_Emergency_Departments_Services.pdf.
[111] Buswell M, Lumbard P, Prothero L, Lee C, Martin S, Fleming J, et al. Unplanned, urgent and emergency care: what are the roles that EMS plays in providing for older people with dementia? An integrative review of policy, professional recommendations and evidence. Emerg Med J 2016;33(1):61–70.
[112] Voss S, Black S, Brandling J, et al. Home or hospital for people with dementia and one or more other multimorbidities: what is the potential to reduce avoidable emergency admissions? The HOMEWARD project protocol. BMJ Open 2017;7(4):e016651. https://doi.org/10.1136/bmjopen-2017-016651.

CHAPTER 6

Change in informal support: creating a caregiving system

Melanie Deist, Abraham P. Greeff
Department of Psychology, Stellenbosch University, Stellenbosch, South Africa

Introduction

> *She gave us everything. Out of love for what she has done, for all the love she gave us, it's now our turn to pay her back. We had a lot of support growing up and now we want to support her.*
>
> ***{Participant 42, female, age 57 years}.***

Caregiving is a complex, multidimensional concept that includes both direct tasks (e.g., hygienic care; meal preparation; health care; transportation; shopping) and indirect responsibilities (e.g., financial management; delegation and management of activities; institutionalization). The growing cost of health care and changing demographic patterns have shifted the responsibility of direct care from formal institutions to informal care partners [1]. When diagnosed with a degenerative disease, such as dementia, reaching a stage where one can no longer effectively take care of oneself is inevitable. In most cases, family members—mostly the spouses or adult children of the person living with dementia—step in as care partners [1–7]. This transition holds several challenges that affect the person living with dementia, the care partner, and the family unit. These families are faced with tasks that are physically exhausting and psychologically distressing. Nonetheless, many families show resilience and are able to overcome the adversity of the illness. But what difficulties do these families face? What factors help these families to adapt to their changing circumstances and counteract the difficulties that impair family functioning? This chapter aims to answer these questions by exploring resilience factors families use to rise above the hardships faced when caring for somebody diagnosed with dementia.

Evidence-informed Approaches for Managing Dementia Transitions
ISBN 978-0-12-817566-8
https://doi.org/10.1016/B978-0-12-817566-8.00006-1

Methods

A literature search on resilience factors in families caring for a person living with dementia was conducted via the electronic databases PubMed, Web of Science, EBSCOhost (Academic Search Premier and PsycARTICLES), and ProQuest. These results were compared with family resilience factors identified in a South African study by Deist and Greeff [8]. These results will be structured around McCubbin and McCubbin's "Resiliency Model of Family Stress, Adjustment, and Adaptation" [9]—further on referred to as "Resiliency Model"—and Walsh's "Resilience Framework" [10,11].

Deist and Greeff [8] focused on experiences of adult children living with a parent diagnosed with dementia and explored family resilience factors that enabled these families to cope more effectively with the hardships of dementia care. The study followed a mixed-method approach, and data were collected from a convenience sample (N = 47). The unit of analysis in this study was the family—a group of two or more people linked through kinship, marriage, domestic partnership, or adoption, who reside in the same household [12]. Family representatives were mostly women (n = 38; 81%) and aged between 20 and 64 years (mean = 47.2; standard deviation [SD] = 9.6). The average income of the participants were slightly higher than the national average, with 11% earning between 1400 ZAR and 2800 ZAR ($100 US and $200 US) a month, 26% earning between 2800 ZAR and 7000 ZAR ($200 US and $500 US) a month, 30% earning between 7000 ZAR and 14,000 ZAR ($500 US and $1000 US) a month, and 34% earning more than 14,000 ZAR ($1000 US) per month.

Unfortunately the literature search on dementia care only obtained studies that focused on the individual resilience of dementia care partners. Resilience may take on a different nature at varying levels of analysis [9]. However, individual resilience does have an impact on family resilience, thus justifying the comparisons made in this chapter [8,13–15].

Living with a parent with dementia: family resilience

Quantitative results

In an attempt to identify factors that help families to cope with the difficulties of dementia care, quantitative data were collected via a demographic questionnaire and seven self-report questionnaires. These questionnaires were selected based on previous family resilience research [13–19] and are in accordance with McCubbin and McCubbin's (1996) Resiliency Model [9]. Family adaptation was measured using the Family Attachment Changeability Index 8 (FACI8) [20]. The independent variables were measured via the Family Hardiness Index (FHI) [21]; the Family Problem Solving and Communication Scale (FPSC) [22]; the Family Time and Routine Index (FTRI) [23]; the Relative and Friend Support Index (RFS) [24]; the Social Support Index (SSI) [25]; and the Family Crisis Oriented Personal Evaluation Scale (F-COPES) [26].

Spearman correlations were calculated to determine the relationship between family adaptation and several demographic variables. Household income ($r = 0.37$; $p = 0.01$) and the number of adults in the household ($r = 0.33$; $p = 0.03$) were the only demographic variables that had a significant positive correlation with family adaptation.

Independent variables that had a significant relationship with family adaptation were determined by calculating Pearson's product–moment correlations. Family resilience scores tended to be higher in families who possessed the following resilience characteristics: family hardiness ($r = 0.71$; $p < 0.01$); family commitment ($r = 0.59$; $p < 0.01$); a sense of family togetherness, which included spending time together ($r = 0.49$; $p < 0.01$); a sense of being in control ($r = 0.63$; $p < 0.01$); an ability to accept problematic issues and minimizing reactivity ($r = 0.49$: $p < 0.01$); an ability to see crises as challenges ($r = 0.56$; $p < 0.01$); community support and a willingness to utilize community resources ($r = 0.35$; $p = 0.02$); and effective problem-solving communication patterns ($r = 0.73$; $p < 0.01$), which encouraged affirming communication ($r = 0.72$; $p < 0.01$); and minimized incendiary communication ($r = -0.67$; $p < 0.01$).

Best-subset multiple regression analyses were conducted to determine the combination of independent variables that best predicted the level of family adaptation in families caring for a parent living with dementia. A multiple R^2 value of 0.56 ($F(5,38) = 10.48$; $p < 0.01$) was obtained. Although the family's overall problem-solving communication pattern was the only significant predictor of family adaptation [$t(41) = 4.82$; $p < 0.01$], passive appraisal of a situation was also present in all of the 20 best subsets, and reframing of a crisis (a coping style) was present in 15 of the 20 best subsets. This implies that passive appraisal and reframing a situation could both still be deemed important variables in the prediction of variance in family adaptation.

Qualitative results

Qualitative data were collected via semistructured interviews with family representatives (*n* = *21*), and the data were analyzed via thematic content analysis. A summary of themes that were identified is depicted in Table 6.1.

Table 6.1 Resilience resources as identified in qualitative interviews (*n* = 21).

Themes	Frequency	%
Internal family characteristics	**f**	**%**
Actively maintaining a positive attitude (e.g., humor, personal time to escape from responsibilities, facing the illness 1 day at a time)	18	86
Acceptance	9	43
Spirituality and religion (e.g., prayer, Bible study)	10	48
Family connectedness	13	62
Good current relationship between family members	12	57
Good past relationship between family members	7	33
Spending time together	6	29
Positive communication patterns	16	76
Open, clear communication	11	52
Patience, avoiding negative communication patterns	11	52
Flexibility	3	14
Consistency (e.g., constant family routines)	3	14
Characteristics of individual family members	9	43
Social support	**21**	**100**
Support from family members	19	90
Support from friends	5	24
Support from others facing similar problems (e.g., support groups)	7	33
Support from religious institution (e.g., church, prayer groups)	1	5
Other community support (e.g., neighbors, community groups)	3	14
Hired help (e.g., hired nurses, day care, respite care)	13	62
Support from doctors	6	29
Help (e.g., psychologists, social workers)	2	10
No community support	9	43
Social isolation	7	33
Financial support	–	–
Self-sufficient (do not get financial support)	13	62
Financial support from family members	6	29
Managing the illness	–	–
Managing symptoms	16	76
Information-seeking on dementia	13	62

The impact of a stressor event: taking care of a parent with dementia

Dementia is an illness characterized by a continuous decline in cognitive and physical abilities. As the illness progresses, people living with dementia become more and more dependent on the help of others, ultimately reaching a stage when they are entirely dependent on their care partners [7,27]. These families need to successfully balance taking care of the person living with dementia and their own household. They are faced with unexpected physical, psychological, financial, and social burdens that test the stability of family systems and coping resources [3,6,28–30]. However, resilient families use resilience resources available to them to buffer against the negative effects of burden, thus optimizing family functioning [4,28] and improving the quality of life for all household members.

Physical and psychological burden

> *I did not know what was waiting for me. In my wildest dreams I never thought that my mom would eat poop; that she would actually give my dad the dog's food to eat right out of the dog's food bowl. They forget things — where they live, who they were, who you are. That is what you have to realise: the disease changes them. They are no longer the people you knew.*
>
> ***{Participant 41, female, age 54 years}.***

When a person living with dementia moves in with their adult offspring (and care partner), family roles slowly begin to change. Initially, there are gradual and subtle shifts in responsibility, but as dementia progresses, the person living with dementia becomes less independent. The care partner initially takes over instrumental tasks, such as cooking, domestic work, financial management, and driving [5,31–34]. However, as the person living with dementia enters later stages of the illness, care partners are often forced to take on a parental role [3,27,28,30] in which the person living with dementia is completely dependent on them for activities of daily living (e.g., dressing, bathing, toileting, and healthcare management) [6,29,30,35,36].

Care partners are often overwhelmed with mounting responsibilities and the increasing dependence of the person living with dementia [29–31,37,38]. They also face emotional burden. Care partners may struggle with feelings of loss, grief, and guilt because of the deterioration of their relationship with the person living with dementia [2,6,27,30,31,36,39–43]. This emotional burden is further exacerbated by anticipatory grief, in which grieving a loved one's inevitable death starts while the person is still alive [27,36,39–41]. Later stages of dementia also bring about cognitive and behavioral changes [3,6,27,29,30,36,41,43–47], which could lead to negative caregiving responses, such as anger, shame, resentment, and regret [27–32,48].

If care partners are unable to adapt by creating a caregiving system that optimizes family functioning, the burdens of dementia care could manifest through psychological problems, such as depression and anxiety [5,6,28,29,34,42,49–51]. The demanding nature of dementia care and the prolonged psychological distress that care partners and their families face could also increase their risk for physical health problems [6,7,29,49,52]. These problems include higher levels of chronic conditions (e.g., diabetes, hypertension, arthritis, ulcers, and anemia) [3], reduced immune responses [3,6], higher rates of symptoms related to stress and physical exertion (e.g., chronic fatigue, headaches) [42], and higher rates of symptoms associated with cardiac or respiratory disease [6,42]. These individuals may also show lower self-rated health behaviors, such as healthy eating habits, getting enough rest, regular exercise, and taking care of their own health problems [3,6,37,53].

Financial burden

> *I have to stay away from work for a whole day to get him to the hospital for his doctor's appointment. That's money I lose. I have to give money out to get here; I have to pay bus fare for both of us just to get here.*
>
> ***{Participant 1, male, age 47 years}.***

Dementia care can be expensive. In some countries, families often need to spend a generous portion of their income on health care for the person living with dementia. These expenses include medical evaluations, consultations at health practitioners, pharmaceuticals, and respite care (e.g., nursing care). The costs of dementia care are not the only expenses these families have to worry about. In most cases, care partners are older individuals who need to take care of their own health as well. In addition, they often still provide financial support to their own children [3,54,55]. Job-related difficulties caused by caregiving responsibilities (e.g., fewer work hours; being overly tired at work; taking unpaid leave; and missing new job opportunities) could also place an economic burden on the family [3,6,7,29,50,55].

Social burden

> *You know, I think that's what it is — I'm not able to socialize with people. Sometimes you just want to talk to somebody and there is nobody that you can talk to. And it depresses you. I sometimes cry. I do have a group of friends, but {they} don't come around anymore.*
>
> ***{Participant 20, female, age 57 years}.***

Care partners are often faced with social isolation [3,6,7,27,29,34,39,42,48,51,56–58]. The impact of social isolation was emphasized in the qualitative interviews by Deist and Greeff [8]. The participants explained that their caregiving responsibilities made it very difficult to leave the house for socialization. Care partners thus withdrew from activities and interests and increasingly stayed at home.

Internal social support and intrafamily relationships are also affected by dementia care. Care partners of people living with dementia often need to execute their caregiving duties while maintaining a healthy relationship with their own spouse and children [55]. Caregiving responsibilities often limit time spent with family [6,29,34,55]. Conflict within the household can arise because of disagreements regarding the division of responsibilities or the implementation of caregiving tasks [6].

Relationships between dementia care partners and their extended family are also put to the test when caring for a person living with dementia. Family conflict often revolves around lack of communication, unequal distribution of caregiving responsibilities, lack of financial support, accusations of neglect or exploitation, differences in opinions regarding the best interests of the person living with dementia, and negative criticism regarding the caregiving process. Tension caused by the burdens of dementia care can expose preexisting cracks in family solidarity, reawaken sibling rivalries, or cause new divisions in family groups [28,29,31,36,59].

What is resilience?

> *{Resilience is} the ability to manage the negative impact of adversity effectively and to transform it into a positive learning experience, thus sustaining physical, psychological and spiritual well-being and reducing vulnerability to future stressors [60].*

When someone living with dementia joins the household of their adult child, the caregiving burdens they face could have a disruptive effect on the functioning of the family system. This transition does not only affect the care partner and person living with dementia but may also influence all the family members. The family resilience process enables families to maintain healthy family functioning, regardless of the disruptive nature of dementia care. Even if family members struggle with the transition initially, resilient families are able to "bounce back" from hardship by invoking recovery factors that promote adaptation and reduce the risk of dysfunction [9,11,61,62].

In McCubbin and McCubbin's Resiliency Model [9], the family resilience process is divided into the adjustment phase and the adaptation phase. During the adjustment phase, families make minor changes in their internal structures in an attempt to manage the burden of a stressor event. The level of adjustment achieved is dependent on several interacting components: the family's appraisal of the stressor, the availability of established patterns of functioning, the family's vulnerabilities, the effectiveness of the family unit's problem-solving and coping skills, and the family's accessibility to resistance resources. If the demands of a stressor are too severe and a family's efforts to adjust are inadequate, a state of maladjustment develops. This typically results in a family crisis. When faced with a family crisis, families need to implement significant changes in their internal structures and patterns of functioning to deal with the stressor at hand. They enter the adaptation phase, which is influenced by the pile up of demands; the family's vulnerabilities; the family's resources; the family's network of social support; coping and problem-solving strategies; and culture and ethnicity. To optimize adaptation, families need to modify family schemas and family paradigms that could influence family coherence. Positive adaptation is achieved if the demands of a stressor event are successfully integrated into the family's established patterns of functioning and balance and harmony are restored [9].

Walsh's Resilience Framework [10,11] combines both ecological and developmental perspectives to explain the nature of family resilience. The framework simplifies the family resilience process by isolating individual, family, and community features that interact with each other to shape family behavior. Walsh [10,11] identified three family domains that were common to most high-functioning families: (1) family belief systems; (2) organizational patterns; and (3) communication [10,11]. Further on, this chapter is structured around these domains to better explain the resilience factors used in families caring for a parent living with dementia.

Family belief systems

Family resilience is fostered by shared family beliefs, which mainly comprise the family's shared values, attitudes, convictions, assumptions, and biases. These beliefs help families make meaning of adversity, encourage optimism, and provide for spiritual support, which in turn aids in healing, growth, and problem-solving [10,11].

Making meaning of adversity

> *I think acceptance is important. It's the cards we've been dealt; nothing can change that. You just have to accept it and deal with it. And you've got to take on a practical attitude.*
>
> ***{Participant 33, female, age 50 years}.***

A family's response to a stressful situation is molded by their subjective appraisal of the stressor event. The meaning the family bestows on the stressor event is influenced by family schema, family paradigms, and the family's sense of coherence. Families need to make sense of unexpected crises before they can effectively restore balance and harmony to the family unit [9—11]. The subjective appraisal of problems faced when caring for a person living with dementia is a better predictor of burden and depression in care partners than the objective severity of the dementia symptoms [4,57].

High caregiving self-efficacy fosters a sense of control over otherwise threatening conditions, which reduces feelings of vulnerability and encourages positive coping strategies [63—66]. The qualitative results of the study by Deist and Greeff [8] revealed that the self-confidence of dementia care partners regarding symptom management played a significant role in family adaptation. Some family representatives believed that adaptation tended to be easier if family members perceived the dementia symptoms as manageable and nondisruptive to their daily lives. Participants managed dementia symptoms through medication or other tried and tested caregiving strategies.

Deist and Greeff [8] identified *reframing of a crisis* as a predictor of family adaptation. This suggested that the family's ability to reframe a situation to make it more manageable copredicts family adaptation. By reframing problems associated with dementia care as a challenge that is comprehensible, manageable, and meaningful to tackle, families are better able to normalize this transition. This fosters confidence that their circumstances will ultimately work out in a favorable way. This meaning-making process influences the family's ability to come to terms with the new norm and betters their competence in selecting adequate coping responses [9,11].

In a dementia caregiving context, care partners have to accept the degenerative nature of the illness if they are to adapt successfully. Studies have shown that the narrative implying that the person living with dementia is still the same as before is often not helpful. The belief that nothing has changed forces expectations on family members to preserve normalcy at any cost. Once families are confronted with the behavioral changes brought about by dementia symptoms, they experience a sense of failure, which evokes feelings of shame and guilt. Families of people living with dementia need support that acknowledges the possibility that their loved one may well be drastically different to the person they previously knew [27,40,41].

Qualitative findings by Deist and Greeff [8] revealed that acceptance played a key role in the resilience process; nearly half of the participants in their study believed that adaptation is only possible once family members are able to accept the condition of the person living with dementia and the effect of the illness on the family unit. This premise was supported by the quantitative results. A significant positive correlation was found between family adaptation and passive appraisal. Furthermore, the best-subset multiple regression analysis identified passive appraisal as a predictor of family adaptation. These results suggest that families tend to experience better family adaptation if they accept their situation and minimize their reactivity toward it.

The meaning-making process is facilitated by the availability of adequate information regarding the causes of dementia, treatment options, and coping resources that could help care partners and their families to adapt. Well-informed family care partners tended to report their caregiving experience more positively, were better able to cope with the problematic behaviors of those living with dementia on a day-to-day basis, were less likely to be depressed [56,57,67], and scored higher in self-reported health behaviors [57]. Active information-seeking is thus a vital component of the family resilience process [27,31,48].

Information-seeking was identified as an effective resilience resource by the majority of participants in the study by Deist and Greeff [8]. For many care partners, internet searches are the initial starting point in their search for answers regarding dementia symptoms [34]. However, some found the internet experience confusing and frustrating with limited success in providing targeted information [44]. Most care partners thus prefer that their primary healthcare providers act as the principal source of health information regarding the dementia diagnosis. They expected assistance regarding information, referrals to support services and other allied health professionals, and guidance in recognizing disease progression [44,68]. Other sources of information used by the participants include structured educational programs, community-based dementia support groups, informational sessions provided by the local government or by nonprofit dementia organizations, pamphlets, articles, and self-help books on dementia, and television and radio programs focusing on this illness [27,34,44,48].

Optimism

> *Stay positive. If you're positive, you will find a way around or you'll make one. But the moment you start being negative I think you plunge everybody into the same hole.*
>
> ***{Participant 33, female, age 50 years}.***

An optimistic view of life is a key family resilience resource that fosters hope for the future and provides families with the strength to rise above adverse circumstances. Unlike families in denial, hopeful families are aware of the realities of a poor prognosis but consciously choose to make the best of the options available to them. Optimism helps families to reframe stressful situations in such a way that they seem manageable. This reinforces confidence in the family's potential to effectively cope with stressor events and fosters a "can do" attitude toward conditions that are beyond their control. Ultimately this minimizes feelings of blame, shame, and guilt [11,62]. Optimism also has a significant impact on coping behaviors, coping effectiveness, and outcome expectancies and is a key factor in the buffering of burden in the family members of persons living with dementia [69].

The most popular family characteristic reported in the study by Deist and Greeff [8] was optimism. Nearly all of the participants actively sought out methods that helped them keep a positive attitude. When faced with a degenerative illness such as dementia, families must accept that the condition of their loved one will only get worse. Family care partners often deal with the poor prognosis of dementia via an emotion-focused coping strategy, in which they try to cope on a day-to-day basis. These families avoid thoughts about the future until it is absolutely necessary. They try to enjoy the time they have left with the person living with dementia [31,32,56,58]. The availability of medication that manages dementia symptoms often buys these families some time with the individual [31]. Spending quality time together and engaging in enjoyable activities as a family is a meaningful way to obtain respite and helps to manage the dementia symptoms [11,31,62]. During this time, families characterized by positivism take joy in the small blessings that shine through [31]. The affirmative focus of optimistic family units motivate care partners to view the caregiving process as a rewarding experience that provides them with a sense of purpose and personal growth [38,48].

The study by Deist and Greeff [8] also identified several other strategies used by participants to stay positive. Some participants used humor to maintain an optimistic view of their situation. Laughter and humor is a healthy form of coping with a grim situation. Humor provides an escape from the burdens associated with dementia caregiving and counteracts depressive symptoms. Laughter can uplift one's spirit, relieve stress, and increase general happiness [62,70]. To ensure that the demands of dementia care do not damper their optimism, several participants also made a point of taking personal time to relax and escape from the responsibilities of caregiving. Family members need to occasionally separate themselves from their caregiving role and focus on their own well-being. Hobbies—such as reading, listening to music, taking a drive, gardening, or going for a walk—helped family members to clear their heads when times were tough [11,34].

Deist and Greeff [8] found a significant positive correlation between family hardiness and family adaptation. Family hardiness facilitates an optimistic family outlook and is an important component of family resilience [9]. Family hardiness is characterized by mutual dependence and the ability to work together as a family unit. Families with high hardiness scores do not view change as a threat but rather as a normal part of life that could bring about opportunities for growth. They participate actively in life events and believe that they have a significant influence on eventual outcomes [19]. Family hardiness is characterized by self-efficacy and a focus on strengths rather than failures. Dementia care partners with higher levels of self-efficacy are better at controlling upsetting thoughts; obtaining respite; and managing dementia-related behavior [65]. Caregiving self-efficacy is also associated with reduced health risks in dementia care partners [71].

Spiritual support

> *I pray. Every morning when I wake up, I ask the Lord to give me strength for the day. I will stay strong if I rely on the Lord. He provides for me. He helps me. So why should I feel miserable? He prepared the road for me. He will not load something on my shoulders that I cannot manage to carry.*
>
> ***{Participant 41, female, age 54 years}.***

Families are continuously faced with the probability that the condition of the person living with dementia might get worse. Spirituality can be viewed as a stable, unchanging rock that anchors care partners when dealing with the ever-changing demands of dementia care. Spiritual beliefs and practices provide families with a purpose beyond themselves and their immediate problems. Families find comfort and hope in their belief that God is ultimately in control. Spirituality buffers the family from the burdens of dementia care and motivates them to continue in the caregiving role. In addition, religious or congregational affiliations connect families to a larger community that provides support and comfort during adverse situations [7,9,11,51,57,62,63,72].

The importance of spirituality in the resilience process was mentioned by half of the participants in the study by Deist and Greeff [8]. Some participants mentioned that they found it difficult to attend spiritual gatherings or congregational activities, but they still practised their spirituality at home and deemed it essential in the adaptation process.

Family organizational patterns

When faced with a crisis event, families sometimes need to modify their rules, family roles, and family patterns of functioning to adapt effectively. These modifications often bring about changes in the organizational patterns of the family. Flexibility, family cohesion and connectedness, social support, and economic resources all have a significant influence on the family's response to these changes [10,11].

Flexibility

> *Things change. Every single day there is something else that is different. Something might help today, but tomorrow you need to find a new way to handle the situation. You just have to adapt. {...} Flexibility is a must.*
>
> ***{Participant 33, female, age 50 years}.***

Families facing a crisis event tend to function best when a sense of balance is achieved between flexibility and structure [11,61,62]. Owing to the progressive nature of dementia, family members often need to take over roles that can no longer be fulfilled by the person living with dementia. Flexibility helps families to view these inevitable changes more optimistically and allows them to adapt to their new sense of normality. However, a lack of rules and routines foster a chaotic family environment characterized by inconsistency, unpredictability, and role confusion, which could ultimately lead to conflict. Daily routines and rituals provide families with feelings of cohesion, comfort, and predictability, which counterbalance the disruptive effects of change [11,62].

Dementia care partners often develop routines for people living with dementia that simplify daily responsibilities. Routines help care partners to deal with dementia-related problems, such as inconsistent sleep patterns, thus reducing caregiving burden and depressive symptoms. These structured routines help both the person and their care partners to feel more competent in their daily tasks and foster a sense of control and stability in their home life [33,56]. Including children and those living with dementia in some household chores can also speed up the completion of daily tasks and thus lessen the burden of caregiving on the care partner. These activities foster a sense of dignity and purposefulness in persons living with dementia. Furthermore, it allows children to participate in the caregiving process and nurtures a sense of pride in them from aiding both the care partner and the person living with dementia. However, some routines might become obsolete as the illness progresses. Care partners need to be flexible enough to allow for new strategies when dealing with new circumstances or when old routines are no longer effective [33,62].

In the qualitative results of Deist and Greeff [8], three participants mentioned routines as an important resilience factor. Flexibility, on the other hand, was also mentioned by three participants. The quantitative results did not provide much support for either of these variables. The only set of routines that was significantly correlated with family adaptation was routines that encourage family members to spend time with each other.

Family cohesion and connectedness

I believe that if you don't love {the person living with dementia}, you will never be able to take care of him. I love {my dad}, my children love him, my husband loves my dad very much — we all love him and have beautiful memories of him — memories of the person he was. I think that helps. I find joy in his happiness.

{Participant 41, female, age 54 years}.

A sense of connectedness and family cohesiveness plays a significant role in the resilience process. When faced with difficult circumstances, families often move away from an individualistic orientation and turn to each other for support. Families need to reach a consensus regarding the commitment style that works best for them—there is no ideal amount of emotional connectedness [11,61,62].

The relationship between family adaptation and connectedness was highlighted in the qualitative and quantitative findings of Deist and Greeff [8]. More than half of the participants identified family connectedness as a key family resilience resource when taking on the role of care partner for a parent living with dementia. Close family relationships provided family members with practical and emotional support when facing the hardships of dementia care. One participant explained that love, mutual respect, and displays of affection between family members motivated them to make the best of their circumstances and comforted them when they felt down. The current relationship between the care partner and the person living with dementia, the past relationship between family members and the individual, and spending time together as a family unit all contributed to family connectedness. These results were supported by the quantitative findings, which revealed a significant positive correlation between family adaptation and the family members' ability to work together and depend on each other in times of hardship.

The relationship quality between care partners and persons living with dementia before onset continues after diagnosis and has a significant influence on the quality of care provided by the care partner. Positive, loving memories of the individual motivate family members to stay positive and often make caregiving less burdensome. Dementia care is often viewed as a token of gratitude through which the family expresses their appreciation for the support they had received from the person living with dementia in the past [58,67]. Good prior relationship quality is related to significantly less burden, better problem-solving skills, more effective communication patterns, a better sense of reward and satisfaction regarding dementia care, less reactivity to memory and behavioral problems, and a higher quality of life for all involved [2,36,73].

Despite the support found for the protective qualities of good prior relationships between care partners and people living with dementia, some studies show a positive correlation between high levels of past closeness and increased distress in care partners. This phenomenon might be due to the loss of closeness between care partners and those living with dementia. Alternatively, it could be a coping strategy care partners use, in which they distance themselves from the person living with dementia. It could also possibly be explained by burnout; care partners with higher levels of connectedness with the person living with dementia are often reluctant to delegate caregiving tasks, thus making it harder to run the caregiving system effectively [2,5].

Current relationships between family members are a central component of family resilience—especially in families where dementia care partners have to take care of their own children as well. The lives of grandchildren change drastically when they live with a cognitively impaired grandparent [46,47,74,75]. Parents serve as gatekeepers for the grandparent–grandchild relationship, and family connectedness inspires emotional closeness between grandchildren and grandparents. This relationship could function as a buffer to the burdens of dementia care. Furthermore, children tend to turn to their parents as a source of support during periods of stress. Close family relationships motivate positive communication between family members. Children also learn positive coping strategies when observing the loving interactions between care partners and the person living with dementia. In turn, children offer care partners emotional and instrumental support, which may alleviate some of the burden associated with dementia care [32,74].

The time a family spends together plays a crucial role in the development of family cohesion. Shared recreation and leisure time between family members promote attachment, happiness, and the enjoyment of shared experiences [47,62,76]. However, activities should be based on the responses of all those participating—including the individual living with dementia. Physical activity, such as family walks or ball games, can be enjoyable for all generations and contains the added benefit of exercise. Singing, dancing, or other music-related activities can reduce tension, evoke memories, and stimulate movement. Spending time together through arts and crafts can foster creativity and provide opportunities for family members to express themselves. Whatever family activity selected, it should always be dignified and gratifying and should focus on the experience itself rather than the end result [47]. The quantitative results of Deist and Greeff [8] further strengthened this premise, showing a significant, positive relationship between routines that encouraged family members to spend time with each other and family adaptation.

Strong levels of attachment between family members often inspire strong feelings of loyalty, reciprocity, and solidarity. This is known as familism and encourages care participants to use the family as a reliable source of support. However, families with high familism often foster rigid views regarding family caregiving, which often lead to dysfunctional thought patterns (e.g., "When a person takes care of a sick relative, he should set aside his interests, and dedicate himself completely to the care of the relative" [77], p. 199]). These thought patterns promote a sense of obligation that increases psychological distress. In addition, intrafamilial conflict may erupt when the behaviors and attitudes of family members do not adhere to these rigid views. Even though a close family relationship has a positive effect on adaptation, it is vital that equilibrium exists between connectedness and separateness to optimize family functioning [77]. Family members must learn to accept each other's individual differences and respect each other's need for separateness and boundaries [11, 61–62].

Social support

> *Emotionally it's a strain, and there's no way you can do it alone. You need someone. My advice to anybody would be to use whatever help you can get. Don't push anyone away. The more hands, the better. Give them one side and you take the other side and together you can carry the burden. But to do it on your own? I don't think so.*
>
> ***{Participant 15, female, age 46 years}.***

When faced with adversity, resilient families have the strength to admit that they need help and are more likely to use different social support structures. These support systems provide families with practical and emotional backing via information, instrumental support services, companionship, respite, and a sense of belonging [9–11,51,62,78]. In families caring for a person living with dementia, social support acts as a buffer against stress and burden and promotes family well-being [57,77]. Care partners and their family members show more positive attitudes toward dementia care if they have adequate social support [79].

In Deist and Greeff's [8] qualitative findings, social support was the only resilience resource identified by all the participants. The sources of social support identified in this study can be divided into three groups: (1) informal support, (2) community support, and (3) professional support.

Informal support

> *It's three of us that help each other and take care of him. During week days, my wife watches him. During the evenings, my sister's daughter helps us out. On weekends it's my turn to watch him. We work as a unit. It makes it easier if you work together as a group.*
>
> ***{Participant 1, male, age 47 years}.***

When dealing with dementia care, informal support from family members and friends is a valuable resilience resource—especially in the early days of diagnosis when formal support is limited. Family members (e.g., siblings, spouses, other extended family members) can provide emotional support and take on a share of the day-to-day responsibilities of dementia care [34,56,57,67].

Talking to friends is also a useful emotional outlet. Care partners with more close, meaningful relationships reported higher levels of overall life satisfaction. Higher levels of social support and social activity also relate to better health outcomes in both dementia care partners and people living with dementia. The support provided by family and friends and thus has a buffering effect on the burdens of dementia care [4,5,51,57,67,77].

Familial support was the most commonly used source of social support identified in the study by Deist and Greeff [8]. Family members provided both emotional and instrumental support. The importance of familial assistance was also hinted at in the demographic results. The number of adults in the household had a significant, positive relationship with family adaptation. Sharing tasks and responsibilities could alleviate the physical strain of dementia care placed on a single family member. Larger families hold the potential for more assistance regarding caregiving tasks, thus possibly promoting family adaptation.

A tendency was noticed in the quantitative results of Deist and Greeff [8] that suggested a relationship between family adaptation and relative and friend support. This finding was, however, not statistically significant.

Community support

> *My husband and I go to a support group. [...] Talking about problems and {caregiving} experiences that upset you helps. Getting those feelings out and talking to a group that knows what you are going through help you to feel good again. And it goes both ways. Some people will now experience behaviours that we already experienced. They might feel like they want to run away, but we can now encourage them.*
>
> ***{Participant 24, female, age 47 years}.***

Community resources could be beneficial to families who lack informal social support. Sources of community support include voluntary organizations, support groups, and religious institutions.

Voluntary dementia organizations often present psychoeducational programs that focus on dementia and dementia home care. These information sessions offer families the opportunity to learn more about the illness and improve their understanding of dementia [3,37,48,56,67]. On occasion, these organizations also offer family-based skills training, which better equips family members with the abilities they need to effectively manage dementia home care. These programs motivate family members to assist the primary care partner with their day-to-day tasks and responsibilities, which in turn buffers the burden of dementia care [80].

Support groups offer the opportunity to share experiences, fears, and uncertainties around dementia care with others that are familiar with the condition. They allow family members of persons living with dementia to lean on the group for emotional support. In return, these individuals give back to others in the group through their emotional presence and helpful recommendations. This helps family members to feel less isolated in their caregiving role [27,48,51,56]. For those uncomfortable with traditional support groups, there exist many online forums and discussion groups that serve a similar purpose. Online support is an alternative means of expression that allows for anonymity. It provides an emotional outlet that is available any time of the day and offers a chance to expand one's friendship network to others with similar experiences [74,81].

Religious or congregational affiliations can also provide emotional support when the burden of dementia care gets too much to handle. Prayer groups and church services promote spirituality, which provides inner strength and buffers against the stresses of dementia care. Church activities also act as a social avenue through which family members of persons living with dementia could take a break and just relax with their peers [10,11].

Deist and Greeff [8] also found that family adaptation tended to be better in families who were better integrated into their community, who more often found support in their community, and who regularly made use of community resources when caring for a family member with dementia. The degree to which families found support in their community and the mobilization of the family to seek and accept help from others were both identified as predictors of family adaptation. However, the contribution of these variables to variance in family adaptation was not statistically significant. It is thus possible that these variables were only included in the best subset of predictors due to chance. The qualitative results of Deist and Greeff [8] showed that the community support used by the participating families were mostly from dementia support groups. Only one participant found support from religious institutions. In contrast, about half of the participants believed that the community provided no support at all.

Professional support

> *First of all, speak to your doctor. Ask your doctor to help you through this like we did. We didn't know anything about this "dementia" thing. We got most of our information at {the hospital}. {We had} a very good doctor there and he always motivated us and told us that we must hang in there.*
>
> ***{Participant 3, female, age 25 years}.***

Professional service providers, such as doctors, psychologists, and social workers, buffer against the burdens of dementia care. Primary healthcare providers are usually the first source of information care partners come in contact with. The families of persons with dementia expect reliable health information when meeting with their doctors and place high value on their recommendations, referrals, or endorsements [8][44][80]. Likewise, mental health practitioners can play a significant role in family adaptation. Counsellors offer family members of persons with dementia a safe space where they can express their emotions without the fear of judgment. Counsellors can help family members to make meaning of negative emotions such as anger, frustration, or resentment and can help them manage their responses to these emotions [27].

Paid respite services, such as adult day care and in-home nursing services, are also helpful community resources that assist families by taking over a portion of the caregiving responsibilities [55][80]. Professional support services minimize the physical burden of dementia care, decrease care-related stress, increase psychological well-being, and lower depression [80]. The relief brought about by hired help was mentioned by more than half of the participants in the study by Deist and Greeff (2015) [8]. Paid help provided these families with instrumental support via cooking, cleaning, and the running of household errands, allowed for employment opportunities for primary care partners, offered an escape from feelings of stress and exasperation, and, in the case of adult day care, provided stimulation and socialization to the person living with dementia.

Economic resources

> *Financially we cope and it is not a burden. My dad is on pension and we use his pension for whatever he needs. If more money is needed, obviously we contribute as a family. We don't need a back-up plan for finances. We as a family can take care of him financially.*
>
> ***{Participant 14, male, age 47 years}.***

Financial security plays a significant role in family adaptation. An adequate economic status and good decision-making skills regarding financial management can improve family resilience. Families dealing with insufficient economic resources potentially have to confront a pile up of stressors connected with poverty, such as unemployment, a lack of health care, substandard housing, violence, crime, and substance abuse. This pile up of stressors could have a negative effect on the emotional well-being of these families [11,62].

The importance of financial stability was apparent in the quantitative findings of Deist and Greeff's [8], which revealed a significant positive correlation between household income and family adaptation. This relationship implied that families with a higher household income tend to be better adapted than families with lower household incomes. Some participants mentioned that troubles at work and financial worries made it more difficult to adapt to the situation. One participant elaborated on this fact, explaining that the costs of treatment and medication put immense financial strain on the family. However, most participants felt that they were doing fine without any financial support, despite the financial burden of dementia care. The families who made use of financial support got financial backing from other family members.

Communication and problem-solving processes

> *We never really argue about it. If we need to make a decision {regarding dementia care}, I discuss it with {my husband} first. I don't decide alone, we decide together. It has to be fine with him too. When I come back from dementia talks, I tell him what I've learned. I include him. We talk. You need to talk {to each other}.*
>
> ***{Participant 15, female, age 46 years}.***

Deist and Greeff [8] found a strong relationship between communication and family adaptation. More than half of the participants identified positive communication patterns as a resilience resource. These participants believed that family adaptation was easier if family members were open about the prognosis of the illness and shared their experiences and knowledge with each other. Sharing emotional experiences fostered an understanding between family members as they were aware of each other's feelings and thus better able to support each other on an emotional level. Participants acknowledged that the nature of the illness tested one's patience and that it was common to become frustrated or lose one's temper when subjected to the continuous burdens associated with this illness. However, the participants emphasized that negative communication patterns, such as shouting and screaming, were ineffective ways of dealing with these problems and that they were better able to cope with a situation once they had calmed themselves.

These findings were mirrored in the quantitative results of Deist and Greeff [8]. Family adaptation had a strong significant positive correlation with family problem-solving communication patterns (affirming communication), while negative, inflammatory communication patterns had a strong significant negative correlation with family adaptation. Furthermore, the best-subset regression analysis identified the family's overall problem-solving communication patterns as the independent variable that best predicted variation in family adaptation.

When dealing with dementia care, families need to discuss how they can collectively cope with the progression of the illness. This means that family members will have to share information regarding the prognosis of the person living with dementia, discuss questions or concerns raised by other family members, and keep each other informed as the situation develops. Secrecy makes it harder for family members to understand and master the situation and ultimately leads to confusion, mistrust and intrafamily conflict. Clear, concise communication encourages collaborative problem-solving within a family unit. Members of well-functioning families openly express their own ideas and encourage all other family members to convey their opinions as well. They tend to accept differences of opinion and avoid criticism, blame, and withdrawal during negotiations on problem resolution. During collaborative problem-solving, resilient families tend to build on small successes and use failure as a learning experience, thus not only reacting to the problem at hand but also preparing the family for future adversities [5,11,61,80].

The emotional support between family members is crucial when dealing with an adverse condition, such as dementia care. Positive, affective communication patterns that include the expression of fondness and love can buffer against emotional burden. Open and emotional sharing allows family members to better understand each other and often strengthens their relationship. However, resilient families are not immune to negative emotions. Conflict is likely to erupt when emotions are intense. Nonetheless, a resilient family should be able to successfully resolve differences of opinion without using inflammatory communication patterns. Well-functioning families are tolerant of emotional outbursts, attempt to understand and respect each other despite differences, and ultimately comfort one another when faced with adversity [5,9,10,36,61].

Positive communication plays an important role in the relationships of care partners and their children. Care partners should be aware that attitudes and opinions communicated within the household have a significant effect on their children's state of mind. A healthy parent–child relationship characterized by open, honest communication should be encouraged. Sharing age-appropriate information regarding dementia is crucial in helping children to understand the condition of a person living with dementia. Even though finding the right words can be difficult, it is important that parents explain to children what is happening using simple language [47,75]. Parents should also reassure children that it is normal to experience a full range of emotions (including sadness, anger, frustration, shame and guilt) when dealing with the burdens of dementia care and encourage them to ask for help when they need it. Children are more likely to ask for assistance if they witness their parents leaning on friends and other adult family members for emotional support. It is important, however, that parents not use their children as an outlet for negative emotions, as this could put more emotional strain on them [74].

Communication also plays a significant role in maintaining the dignity of the person living with dementia. During the early stages of dementia, individuals often experience a deep sense of loss. They are aware of their deteriorating capabilities and are often fearful of the manifestations of the disease. Care partners should validate these emotional expressions through empathetic listening and expressions of support [27]. A sense of dignity can only be maintained if the person is treated like an adult; this means including them in conversations, valuing their opinions, listening to them, attending to their needs, and offering them dignified explanations when they are confused [33,82]. A person living with dementia who progress to the next stage of the disease can also at times feel a vague sense of wrongness that could lead to anxiety and depression. Care partners should communicate affection, reassurance, and remembrance to reaffirm their relationship with the person. When appropriate, expressive approaches using physical touch (e.g., hugs) work particularly well [27].

Culture

> *I think it's expected of us in the society that we live in. We need to be there for Dad. That's just the way we were brought up. I don't know of any other way. It has always been like this. I remember my grandma stayed with us when she got old and my parents took care of her. She also became old and frail and that is just the norm. So now it's our turn. You just carry on with what needs to be done and accept it.*
>
> ***{Participant 5, male, age 42 years}.***

The pathway to family adaptation is unique for each family, making it difficult to discover a one-size-fits-all blueprint for family resilience. No single model fits all families or situations, and family resilience should be measured with regard to the family's sociocultural context. When exploring the impact of dementia on care partners and their families, the structure, values, perspectives, resources, and life challenges faced by the family should be taken into consideration [9,11].

Culture has a significant influence on perceptions, appraisal, and coping responses. This infers that the sociocultural background of care partners could dictate their beliefs regarding dementia, their perceptions regarding caregiving, the coping strategies they use to manage crisis situations, and their willingness to seek social support [54,63,83].

Filial piety, for example, is a family-centered cultural value that places a societal expectation on adult children to take care of their ill parents. Meeting these expectations provides a sense of fulfillment and often acts as a protective factor against caregiver burden. These same beliefs, however, demonize external support and view it as abandonment of one's caregiving responsibilities. Care partners with a high sense of filial obligation and thus reject a resilience resource that could ease their burden [36,67,76,84–86]. Western society, on the other hand, relies more on parent–child connectedness as motivation for taking on a caregiving role. The adult children of persons living with dementia are happy to take care of their parent as a gesture of thanks for the love they received in the past [40,87]. However, Western societies often pushes a narrative that overemphasizes the view that the person living with dementia is "still the same person as before," which could foster ignorance regarding possible behavioral changes connected to the illness. Ignorance regarding the probability of behavioral changes could cause great emotional burden for care partners that blame themselves or the individual when these symptoms manifest [40,41,84].

Some individuals are unfamiliar with Western biomedical concepts, such as dementia. Public ignorance regarding the illness, the behavioral abnormality of people living with dementia, and the possibility of a familial predisposition to dementia expose individuals and their families to stigmatization. Some believe that the disease is contagious, thus causing fear and misunderstanding between the public and the people with dementia. Others believe dementia is a form of insanity, which is usually attributed to the individual's personal character or the failure of their families to take proper care of them. On the other end of the continuum is normalization—the view that dementia is a nonfatal, normal part of the aging process rather than a disease. In these societies, the older population is divided into groups: those who "aged well" and those who did not. People living with dementia fall into the latter group, undergoing a less desired trajectory of aging. Such societies distance themselves from these individuals, perhaps seeing them as childish, foolish, or confused. Stigmatization devalues both persons living with dementia and their families. It causes humiliation and shame and is a core reason for the social withdrawal and isolation experienced by many of these families [3,35,45,88,89].

Certain groups, such as women [34,42,90,91], the LGBTQ+ community [3], ethnic minorities [5,63,86], or people with younger-onset dementia [3,32,41,56], may experience additional challenges on top of those associated with dementia care. These families might face racism, discrimination, and other forms of stigmatization. The unique experiences of these groups should thus be taken into consideration when exploring their family functioning and resilience resources.

Conclusion

The responsibility of care for older and aging family members is often placed on the shoulders of adult children, who also have to balance their careers, spousal relationships, and child rearing. Taking on the role of primary care partner for a person living with dementia can severely disrupt the balance and harmony of a family unit. Families need to use the family resilience resources available to them to navigate through this transition. It builds on family strengths and encourages forward-looking family views that focus on success rather than failure. It increases optimism, fosters acceptance, and overall lightens the burden of dementia care.

In this chapter, several family resources were discussed that facilitated the resilience process in families caring for a parent diagnosed with dementia in their home. This information could be used in the development of intervention programs that help these families to create an environment that maximizes adaptation. Skill development interventions could focus on positivism, coping strategies, communication, emotional intelligence development, time management, financial management, and the effective management of dementia symptoms to help families acquire an internal skillset that facilitates the resilience process.

Familial support is also a crucial family resilience factor in a population group plagued by social isolation. A better understanding of dementia and its effects on care partners can be achieved through culturally sensitive education programs. Family-based education on the nature, progress, and associated burdens of dementia could foster better understanding between family members and might even elicit more support from relatives.

However, family resilience is not limited to the internal characteristics of a family unit. The adaptation process could be hindered by societal views characterized by apathy and stigma. Education campaigns and dementia drives could raise public awareness of the illness, thus gradually decreasing stigma. Drawing attention to available services, how these services can be accessed, and how it could improve the lives of care partners could also normalize these social resources, thus allowing care partners and their families to use these services guilt free. External support can also be stimulated through state health policies. Access to affordable medical care and the availability of state-run respite services could alleviate some of the stress experienced by dementia care partners. Cultural sensitivity of primary healthcare practitioners should also be a priority.

Given the projected increase in the prevalence of dementia, and the fact that most dementia patients receive home care by family members, there is a need for intervention strategies that help families cope with the demands of dementia care. Supporting informal care partners helps to avoid burnout, reduces physical and mental health problems, minimizes potentially harmful behavior such as elder abuse, and ultimately motivates family members to maintain their caregiving duties.

References

[1] Swanson EA, Jensen DP, Specht J, Johnson ML, Maas M, Saylor D. Caregiving: concept analysis and outcomes. Sch Inq Nurs Pract Int J 1997;11(1):65–76.
[2] Ablitt A, Jones GV, Muers J. Living with dementia: a systematic review of the influence of relationship factors. Aging Ment Health 2009;13(4):497–511.
[3] Brodaty H, Donkin M. Family carers of people with dementia. Dementia. 4th ed. CRC Press; 2010. p. 137–52.
[4] Chiou CJ, Chang H-Y, Chen IP, Wang HH. Social support and caregiving circumstances as predictors of caregiver burden in Taiwan. Arch Gerontol Geriatr 2009;48(3):419–24.
[5] Mitrani VB, Czaja SJ. Family-based therapy for dementia caregivers: clinical observations. Aging Ment Health 2000;4(3):200–9.
[6] Schulz R, Martire LM. Family caregiving of persons with dementia: prevalence, health effects, and support strategies. Am J Geriatr Psychiatry 2004;12(3):240–9.
[7] Wright SD, Pratt CC, Schmall VL. Spiritual support for caregivers of dementia patients. J Relig Health 1985;24(1):31–8.
[8] Deist M, Greeff AP. Living with a parent with dementia: a family resilience study. Dementia 2016;16(1):126–41.

[9] McCubbin HI, Mccubbin MA. Resiliency in families: a conceptual model of family adjustment and adaptation in response to stress and crises. In: Family assessment: resiliency, coping and adaptation – inventories for research and practice. University of Wisconsin System, University of Wisconsin System; 1996. p. 1–64.
[10] Walsh F. A family resilience framework: innovative practice applications. Fam Relat 2002;51(2):130–7.
[11] Walsh F. Family resilience: a framework for clinical practice. Fam Process 2003;42(1):1–18.
[12] Nam CB. The concept of the family: demographic and genealogical perspectives. Soc. Today 2004;2(2):1–5.
[13] Greeff AP, De Villiers M. Optimism in family resilience. Soc. Work Pract. Res. 2008;20(1):21–34.
[14] Greeff AP, Thiel C. Resilience in families of husbands with prostate cancer. Educ Gerontol 2012;38(3):179–89.
[15] Jonker L, Greeff AP. Resilience factors in families living with people with mental illnesses. J Community Psychol 2009;37(7):859–73.
[16] Bishop M, Greeff AP. Resilience in families in which a member has been diagnosed with schizophrenia. J Psychiatr Ment Health Nurs 2015;22(7):463–71.
[17] Greeff AP, Joubert A-M. Spirituality and resilience in families in which a parent has died. Psychol Rep 2007;100(3):897–900.
[18] Greeff AP, Wentworth A. Resilience in families that have experienced heart-related trauma. Curr Psychol 2009;28(4):302–14.
[19] Greeff AP, Vansteenwegen A, Ide M. Resiliency in families with a member with a psychological disorder. Am J Fam Ther 2006;34(4):285–300.
[20] McCubbin HI. Family attachment and changeability index 8 (FACI8). In: Family stress, coping and health project, school of human ecology, 1300 Linden Drive. Madison, WI: University of Wisconsin-Madison; 2002. 53706-1575.
[21] McCubbin MA, Mccubbin HI, Thompson AI. FHI: family hardiness index. Family assessment: resiliency, coping and adaptation – inventories for research and practice. Madison, Wisconsin: University of Wisconsin System; 1996. p. 239–305.
[22] McCubbin HI. Family problem solving communication (FPSC). Family stress, coping and health project, School of Human Ecology, 1300 Linden Drive. Madison, WI: University of Wisconsin-Madison; 2002. 53706-1575.
[23] McCubbin HI. Family time and routines Index (FTRI). Family stress, coping and health project, School of Human Ecology, 1300 Linden Drive. Madison, WI: University of Wisconsin-Madison; 2002. 53706-1575.
[24] McCubbin HI, Thompson AI, McCubbin MA. Family assessment: resiliency, coping and adaptation: inventories for research and practice. Madison, Wis: University of Wisconsin Publishers; 1996.
[25] McCubbin HI, Patterson JM, Glynn T. SSI: social support index. In: Family assessment: resiliency, coping and adaptation – inventories for research and practice. Madison, Wisconsin: University of Wisconsin System; 1996. p. 357–89.
[26] McCubbin HI, Olson D, Larson A. F-COPES: family crisis oriented personal evaluation scales. In: Family assessment: resiliency, coping and adaptation – inventories for research and practice. Madison, Wi: University of Wisconsin System; 1996. p. 455–507.
[27] Doka KJ. Grief, multiple loss and dementia. Bereave Care 2010;29(3):15–20.
[28] Tremont G, Davis JD, Bishop DS. Unique contribution of family functioning in caregivers of patients with mild to moderate dementia. Dement Geriatr Cognit Disord 2006;21(3):170–4.

[29] Aguglia E, Onor ML, Trevisiol M, Negro C, Saina M, Maso E. Stress in the caregivers of Alzheimer's patients: an experimental investigation in Italy. Am J Alzheimer's Dis Other Dementias 2004;19(4):248–52.
[30] Kjällman-Alm A, Norbergh K-G, Hellzen O. What it means to be an adult child of a person with dementia. Int J Qual Stud Health Well-Being 2013;8(1):21676.
[31] Adams KB. The transition to caregiving. J Gerontol Soc Work 2006;47(3–4):3–29.
[32] Nichols KR, Fam D, Cook C, Pearce M, Elliot G, Baago S, et al. When dementia is in the house: needs assessment survey for young caregivers. Can J Neurol Sci 2013;40(01):21–8.
[33] Askham J, Briggs K, Norman IAN, Redfern S. Care at home for people with dementia: as in a total institution? Ageing Soc 2006;27(1):3–24.
[34] Grigorovich A, Rittenberg N, Dick T, McCann A, Abbott A, Kmielauskas A, et al. Roles and coping strategies of sons caring for a parent with dementia. Am J Occup Ther 2015;70(1). 7001260020p1.
[35] Werner P, Goldstein D, Buchbinder E. Subjective experience of family stigma as reported by children of alzheimer's disease patients. Qual Health Res 2010;20(2):159–69.
[36] Podgorski C, King DA. Losing function, staying connected: family dynamics in provision of care for people with dementia. Generations 2009;33(1):24–9.
[37] Tatangelo G, McCabe M, Macleod A, You E. "I just don't focus on my needs." the unmet health needs of partner and offspring caregivers of people with dementia: a qualitative study. Int J Nurs Stud 2018;77:8–14.
[38] Sanders S. Is the glass half empty or half full? Soc Work Health Care 2005;40(3):57–73.
[39] Holley CK, Mast BT. The impact of anticipatory grief on caregiver burden in dementia caregivers. Gerontol 2009;49(3):388–96.
[40] Sikes P, Hall M. "It was then that I thought 'whaat? This is not my Dad": the implications of the 'still the same person' narrative for children and young people who have a parent with dementia. Dementia 2016;17(2):180–98.
[41] Hall M, Sikes P. "It would Be easier if she'd died": young people with parents with dementia articulating inadmissible stories. Qual Health Res 2017;27(8):1203–14.
[42] Baumgarten M, Battista RN, Infante-Rivard C, Hanley JA, Becker R, Gauthier S. The psychological and physical health of family members caring for an elderly person with dementia. J Clin Epidemiol 1992;45(1):61–70.
[43] Furlini L. The parent they knew and the "new" parent: daughters' perceptions of dementia of the alzheimer's type. Home Health Care Serv Q 2001;20(1):21–38.
[44] Peterson K, Hahn H, Lee AJ, Madison CA, Atri A. In the Information Age, do dementia caregivers get the information they need? Semi-structured interviews to determine informal caregivers' education needs, barriers, and preferences. BMC Geriatr 2016;16(1).
[45] Navab E, Negarandeh R, Peyrovi H, Navab P. Stigma among Iranian family caregivers of patients with Alzheimer's disease: a hermeneutic study. Nurs Health Sci 2012;15(2):201–6.
[46] Fuh JL, Wang SJ, Juang KD. Understanding of senile dementia by children and adolescents: why grandma can't remember me? Acta Neurol Taiwanica 2005;14(3):138–42.
[47] Winters S. Alzheimer disease from a child's perspective. Geriatr Nurs 2003;24(1):36–9.
[48] Bekhet AK, Avery JS. Resilience from the perspectives of caregivers of persons with dementia. Arch Psychiatr Nurs 2018;32(1):19–23.
[49] Pinquart M, Sörensen S. Differences between caregivers and noncaregivers in psychological health and physical health: a meta-analysis. Psychol Aging 2003;18(2):250–67.
[50] Goren A, Montgomery W, Kahle-Wrobleski K, Nakamura T, Ueda K. Impact of caring for persons with Alzheimer's disease or dementia on caregivers' health outcomes: findings from a community based survey in Japan. BMC Geriatr 2016;16(1).

[51] Sanders S, Ott CH, Kelber ST, Noonan P. The experience of high levels of grief in caregivers of persons with alzheimer's disease and related dementia. Death Stud 2008;32(6):495–523.
[52] Abdollahpour I, Nedjat S, Noroozian M, Salimi Y, Majdzadeh R. Caregiver burden. J Geriatr Psychiatry Neurol 2014;27(3):172–80.
[53] Sakurai S, Onishi J, Hirai M. Impaired autonomic nervous system activity during sleep in family caregivers of ambulatory dementia patients in Japan. Biol Res Nurs 2015;17(1):21–8.
[54] Mayston R, Lloyd-Sherlock P, Gallardo S, Wang H, Huang Y, Montes de Oca V, et al. A journey without maps—understanding the costs of caring for dependent older people in Nigeria, China, Mexico and Peru. PLoS One 2017;12(8):e0182360.
[55] Vreugdenhil A. 'Ageing-in-place': frontline experiences of intergenerational family carers of people with dementia. Health Sociol Rev 2014;23(1):43–52.
[56] Quinn C, Clare L, Pearce A, van Dijkhuizen M. The experience of providing care in the early stages of dementia: an interpretative phenomenological analysis. Aging Ment Health 2008;12(6):769–78.
[57] Haley WE, Levine EG, Brown SL, Bartolucci AA. Stress, appraisal, coping, and social support as predictors of adaptational outcome among dementia caregivers. Psychol Aging 1987;2(4):323–30.
[58] McDonnell E, Ryan AA. The experience of sons caring for a parent with dementia. Dementia 2013;13(6):788–802.
[59] Peisah C, Brodaty H, Quadrio C. Family conflict in dementia: prodigal sons and black sheep. Int J Geriatr Psychiatry 2006;21(5):485–92.
[60] Grafton E, Gillespie B, Henderson S. Resilience: the power within. Oncol Nurs Forum 2010;37(6):698–705.
[61] Patterson JM. Understanding family resilience. J Clin Psychol 2002;58(3):233–46.
[62] Black K, Lobo MA. Conceptual review of family resilience factors. J Fam Nurs 2008;14(1):33–55.
[63] Haley WE, Roth DL, Coleton MI, Ford GR, West CAC, Colllins RP, et al. Appraisal, coping, and social support as mediators of well-being in Black and White family caregivers of patients with Alzheimer's disease. J Consult Clin Psychol 1996;64(1):121–9.
[64] Gilliam CM, Steffen AM. The relationship between caregiving self-efficacy and depressive symptoms in dementia family caregivers. Aging Ment Health 2006;10(2):79–86.
[65] Au A, Lai M-K, Lau K-M, Pan P-C, Lam L, Thompson L, et al. Social support and well-being in dementia family caregivers: the mediating role of self-efficacy. Aging Ment Health 2009;13(5):761–8.
[66] Gonyea JG, O'Connor M, Carruth A, Boyle PA. Subjective appraisal of Alzheimer's disease caregiving: the role of self-efficacy and depressive symptoms in the experience of burden. Am J Alzheimer's Dis Other Dementias 2005;20(5):273–80.
[67] Yamashita M, Amagai M. Family caregiving in dementia in Japan. Appl Nurs Res 2008;21(4):227–31.
[68] Lampley-Dallas VT, Mold JW, Flori DE. African-American caregivers' expectations of physicians: gaining insights into the key issues of caregivers' concerns. J Natl Black Nurses' Assoc 2005;16(1):18–23.
[69] Gottlieb BH, Rooney JA. Coping effectiveness: determinants and relevance to the mental health and affect of family caregivers of persons with dementia. Aging Ment Health 2004;8(4):364–73.
[70] Tan T, Schneider MA. Humor as a coping strategy for adult-child caregivers of individuals with alzheimer's disease. Geriatr Nurs 2009;30(6):397–408.
[71] Rabinowitz YG, Mausbach BT, Thompson LW, Gallagher-Thompson D. The relationship between self-efficacy and cumulative health risk associated with health behavior patterns in female caregivers of elderly relatives with alzheimer's dementia. J Aging Health 2007;19(6):946–64.

[72] Karlin NJ. An analysis of religiosity and exercise as predictors of support group attendance and caregiver burden while caring for a family member with Alzheimer's disease. J Ment Health Aging 2004;10(2):99–106.
[73] Daire AP. The influence of parental bonding on emotional distress in caregiving sons for a parent with dementia. Gerontol 2002;42(6):766–71.
[74] Celdrán M, Villar F, Triadó C. When grandparents have dementia: effects on their grandchildren's family relationships. J Fam Issues 2012;33(9):1218–39.
[75] Celdrán M, Triadó C, Villar F. My grandparent has dementia. J Appl Gerontol 2010;30(3):332–52.
[76] Pan Y, Jones PS, Winslow BW. The relationship between mutuality, filial piety, and depression in family caregivers in China. J Transcult Nurs 2016;28(5):455–63.
[77] Losada A, Márquez-González M, Knight BG, Yanguas J, Sayegh P, Romero-Moreno R. Psychosocial factors and caregivers' distress: effects of familism and dysfunctional thoughts. Aging Ment Health 2010;14(2):193–202.
[78] Nawi NHM, Megat Ahmad PH, Malek DA, Cosmas G, Ibrahim H, Voo P, et al. Structural relationship between emotional and social support for young adult carers towards intergenerational care of the multi-ethnic elderly. Qual Ageing Older Adults 2017;18(3):188–200.
[79] Johannesen M, LoGiudice D. Elder abuse: a systematic review of risk factors in community-dwelling elders. Age Ageing 2013;42(3):292–8.
[80] Lindsey Davis L. Family conflicts around dementia home-care. Fam Syst Health 1997;15(1):85–98.
[81] McKechnie V, Barker C, Stott J. The effectiveness of an internet support forum for carers of people with dementia: a pre-post cohort study. J Med Internet Res 2014;16(2):e68.
[82] van Gennip IE, Pasman HR, Oosterveld-Vlug MG, Willems DL, Onwuteaka-Philipsen BD. How dementia affects personal dignity: a qualitative study on the perspective of individuals with mild to moderate dementia: table 1. J Gerontol Ser B Psychol Sci Soc Sci 2014;71(3):491–501.
[83] Kosloski K, Schaefer JP, Allwardt D, Montgomery RJV, Karner TX. The role of cultural factors on clients' attitudes toward caregiving, perceptions of service delivery, and service utilization. Home Health Care Serv Q 2002;21(3–4):65–88.
[84] Lee Y-R, Sung K-T. Cultural differences in caregiving motivations for demented parents: Korean caregivers versus American caregivers. Int J Aging Hum Dev 1997;44(2):115–27.
[85] Khalaila R, Litwin H. Does filial piety decrease depression among family caregivers? Aging Ment Health 2011;15(6):679–86.
[86] Cox C, Monk A. Hispanic culture and family care of alzheimer's patients. Health Soc Work 1993;18(2):92–100.
[87] Lee Y-R, Sung K-T. Cultural influences on caregiving burden: cases of Koreans and Americans. Int J Aging Hum Dev 1998;46(2):125–41.
[88] Liu D, Hinton L, Tran C, Hinton D, Barker JC. Reexamining the relationships among dementia, stigma, and aging in immigrant Chinese and Vietnamese family caregivers. J Cross Cult Gerontol 2008;23(3):283–99.
[89] Ayalon L, Areán PA. Knowledge of Alzheimer's disease in four ethnic groups of older adults. Int J Geriatr Psychiatry 2004;19(1):51–7.
[90] Miller B, Cafasso L. Gender differences in caregiving: fact or artifact? Gerontol 1992;32(4):498–507.
[91] Yee JL, Schulz R. Gender differences in psychiatric morbidity among family caregivers. Gerontol 2000;40(2):147–64.

CHAPTER 7

Transitioning from home in the community to an assisted living residence

Lynn McCleary[1], Mackenzie Powell[2], Willian Dullius[3]
[1]Department of Nursing, Brock University, St. Catharines, Ontario, Canada; [2]Quinte Health Care, Belleville, Ontario, Canada; [3]Escola Estadual de Ensino Médio Cônego João Batista Sorg - Government of the Rio Grande do Sul, Carazinho, Rio Grande do Sul, Brazil

Introduction

Much as people want to stay in their homes, for some, there comes a time when this is not possible. Being unable to manage at home and moving are both profoundly meaningful for anyone, perhaps more so for persons with dementia.

Home is more than the place where we live. It is a source of comfort and security, a private refuge where we have freedom and choice [1,2]. Home is the objects, sounds, and smells around us. Activities we do at home and as part of everyday life at home bring us meaning. Home is the familiarity of our neighborhood and the people there [2]. It is associated with self-identity and well-being and with feelings of mastery and empowerment [1,3]. Home is a place of connection and relationships to family and friends [1].

The home environment can promote wellness and personhood for those with dementia. It has been described as an expression of self and a "repository of memories" (p. 31), and a sense of home remains, even in the advanced stages of dementia [3]. Home is connected to aspects of quality of life among people with dementia, including connectedness to the social and physical environment, belonging, relationships, experiencing autonomy and purpose in everyday life, and feeling settled with a sense of place [4].

As dementia progresses, the benefits of living at home in the community may be outweighed by the challenges of managing activities of daily living and household tasks. Home can become more a place of vulnerability than of security [2]. Some are able to manage by enrolling with community-based services or hiring extra help at home. However, for many, these are not feasible or acceptable options. Moving to an assisted living residence

Evidence-informed Approaches for Managing Dementia Transitions
ISBN 978-0-12-817566-8
https://doi.org/10.1016/B978-0-12-817566-8.00007-3

can provide needed support and relief of the burden of household maintenance. Yet, at the same time, moving to a new environment threatens a person with dementia's meaningful existence and sense of being autonomous [2].

This chapter describes the process of transition from a home in the community to life in an assisted living residence. It draws on research about the transition to an assisted living residence, much of which is qualitative and cross-sectional, with some longitudinal qualitative studies and some epidemiologic studies. Findings from the Dementia Transitions Study [5] are included (Chapter 1). Thirteen participants in that study (aged 77–91 years) moved to an assisted living residence. Interviews were conducted with four persons with dementia and 12 care partners (aged 43–81 years). Four care partners were spouses, two of whom moved with the person with dementia.

What is assisted living and who lives there?

There are a variety of terms for assisted living residences. Terminology varies across jurisdictions. In this chapter, we refer to *assisted living residences,* also known as assisted living facilities, residential care homes, care homes, and retirement homes. The most common reason for moving to an assisted living residence is inability to safely live at home independently, in conjunction with lack of resources to support living at home [6,7]. The move is often precipitated by a change in health status or functioning, difficulty with independent activities of daily living, or declining health of a care partner [8]. It may involve relocating to a different region to be closer to children [9].

Assisted living residences can be distinguished from long-term care homes (also known as nursing homes and skilled nursing facilities) with respect to the amount and kinds of care provided, cost to residents, funding models, and underlying philosophy of care. However, there is tremendous diversity among assisted living residences. The assisted living model was developed as an intermediate step between independent living and long-term care homes. It was originally a social model of care, with limited healthcare services. The model espouses a home-like environment, respecting and promoting residents' autonomy and independence. Typically, assisted living residences provide housekeeping, meals, social and recreational activities, and assistance with personal care needs, including dispensing medication. Most have private rooms or units, with some offering a choice of shared rooms. Some are located within continuing care retirement communities that offer independent living, assisted living, and long-term care homes on campus, where residents may move if their care needs change. There is considerable variability in the extent to which assisted living residences are regulated.

Most assisted living residences operate on a private-pay model, with a base rate of payment and additional fees for extra health or personal care services. In 2018, the average monthly base rate in Ontario, Canada, was $3618. It was $3500 in 2014 in the U.S. [10]. Costs vary widely within jurisdictions. In the U.S., assisted living is increasingly being recognized by Medicaid as an alternative to long-term care homes and 41 states have some Medicaid coverage [11]. Access to and choices about assisted living depend on ability to pay [12].

In North America, just over 1% of older persons live in assisted living residences [13,14]. The prevalence of dementia among residents is increasing, with estimates in the U.S. of between 42% and 89% [15,16]. Dementia care units within assisted living residences are increasingly common, but most residents with dementia do not live in these "special care units." The fees in dementia units are about 30% higher than the assisted living base rate [16,17]. The average length of stay is shorter for persons with dementia than for other residents, primarily because they are likely to be discharged and transferred to a long-term care home because of behavioral symptoms requiring more supervision than the assisted living residence can provide or that require more supervision than the resident can afford [18,19]. However, behavioral symptoms are common in assisted living residences; a national U.S. study found that 38% of residents had behavioral symptoms on admission [16].

What do we mean by transition to an assisted living residence?

The transition to an assisted living residence is more than moving homes. It begins with the first thought of moving, through deciding to move, deciding where to move to, moving, settling in to the assisted living residence, and finally, feeling at home there. In addition to the person with dementia, the transition involves care partners, family, friends, and neighbors who may provide assistance with the transition and whose lives are affected by it. It also involves healthcare providers and residence staff.

Evidence about each of these elements of the transition and associated factors is discussed in the remainder of this chapter.

It is not unusual for the transition to assisted living to happen simultaneously with or shortly after other transitions discussed in this book. For example, safety problems that prompt thinking about moving may also be an impetus to seeking medical care [20]. When the diagnosis is made later in the course of the dementia, obtaining medical care may result in loss of a driver's license and a recommendation to move. The move to an assisted living residence may be precipitated by a health decline or hospital admission [3,21]. The stress of the move to an assisted living residence may precipitate acute symptoms of underlying comorbid chronic illness, requiring treatment in hospital [22]. Thus the transition should be considered in the context of other transitions the person may be experiencing.

Deciding to move

For some, the decision to move to an assisted living residence is made quickly, particularly when the move happens after an illness, hospitalization of the person with dementia, or the loss of a care partner [21,23]. For example, Bill, a participant in the Dementia Transitions Study, lived at home with his wife for 7 years after the initial appearance of symptoms and 3 years after being diagnosed with mixed dementia. He attended a respite day program and received home care services. According to their son, when Bill's wife was diagnosed with advanced cancer and awaiting surgery she was *"literally beside herself with worry about what was going to happen to him. Because she couldn't, she couldn't release herself to her own illness until that was resolved."* Bill's sons got involved in seeking alternatives for him. A case manager placement officer, who was the gatekeeper for government-funded assisted living residences and long-term care homes, assessed him as needing assisted living. There was a long waiting list for government-provided assisted living and few options near their community. Overnight respite care was arranged as a temporary solution. The family learned about a privately operated assisted living residence from a friend, and a bed became available there days before Bill was to go to the overnight respite care. Bill's son immediately flew in from out of town, moving Bill and his belongings to his new residence. The placement officer disagreed with the decision. Bill's wife was immensely relieved.

For others, there is a gradual recognition of a need for more assistance than is available or affordable at home. Ambivalence about moving to assisted living is common among persons with dementia. The ambivalence centers on what is being given up, such as independence, freedom to choose what to do and with whom, and engaging in meaningful activities, versus the increasing burden of housekeeping, isolation, loneliness, and risks to safety and well-being [3,24,25]. This ambivalence is illustrated by a person with dementia in the Dementia Transitions Study as he anticipated moving with his wife. While he accepted that they needed to move, he did not want to give up his freedom:

> *Well, it is what we fear it to be. Psychologically, neither one of us is up to it. We don't want to go into living in one room. So, I know that we don't have to live in one room but that is the only spot in the building you can call your own. So, if you wanted to get out of that room, you have to be with other people, whether you want to or not or whether you are compatible or not. You are with people who you haven't got a clue who they are until you moved into that place. Whether these people are friendly or whether they are just looking for a new cover. I don't want nothing to do with them.*

Ambivalence about what is being given up is illustrated by a participant in the longitudinal study by Thein et al. [25], who feared losing independence:

> *You need care and you'd have to do what you were told, you know. You'd have to do all what, you couldn't do what you liked. And you know, you're not like your own boss kind of thing. You'd have somebody over you and so I'm not thinking about it (p. 12).*

The person with dementia's recognition of a need to move is influenced by opinions of relatives and healthcare providers [3,24]. A participant in the study by Aminzadeh et al. [24] explained

> *My daughter says, 'you work far too hard. You worked for eighty years. Let somebody else do it.' ... generally a person like myself moving out of this house and going to an establishment like we're going to uh, you wouldn't be able to cook again and that's cutting a big slice of your life out, you know. Things you enjoyed uh, in one whack, you're not going to do that (p. 491).*

The process of relatives persuading the person with dementia that moving is a good idea can go on for months [26]. Coming to accept the need for the move before it happens may be associated with better adjustment.

Uncertainty about when to move is common [27]. Descriptions by persons with dementia of the process of deciding to move often involve recognizing problems with physical health, mobility, and dependence, which may be identified as age related [3,21,25]. For example, a participant in the study by Aminzadeh et al. [3] described

I know that I got into some jams there for a while. I was dragging my feet, and life didn't look all that exciting. I can't cook, I can't clean. I do nothing at home .... You worry — My gosh, I really should be doing so and so — I wonder if I ordered my gas and paid my phone bills ... So, that when we decided 'Let's go have a look at some [residential care] homes.' It makes life easier, because your mind takes little day trips (p. 32).

For care partners, the most common concerns leading to the decision to move are safety of the person with dementia or the caregiver, related to progression of dementia symptoms, behaviors, and inability to function, sometimes in conjunction with decline in the care partner's health [21,28]. Care partners may feel guilty about their unsuccessful efforts to keep people at home [12]. Similar to persons with dementia, care partners can experience uncertainty and ambivalence as they come to the conclusion that more assistance is needed than can be provided at home.

The process of deciding to move and choosing an assisted living residence involves seeking information, receiving advice, and weighing alternatives. Factors that influence decisions about choice of an assisted living residence include availability and urgency of the move, cost, previous knowledge about the residence, knowing someone who lives there, proximity to the current residence, and proximity to adult children who have caregiving responsibilities for the person with dementia. Sources of information and advice include healthcare providers, such as family physicians or memory specialists, community case managers, commercial websites, and family and friends with similar experiences. Obtaining information during a crisis, such as a move precipitated by loss of a care partner or hospital admission of a person with dementia, may be difficult. Bill's son, who participated in the Dementia Transitions Study explained

... and we walked into this and you have a 1 hour meeting with a placement liaison and she goes over the three levels of care. And she does this two, three times a day with families. And you know, you got the handout as well and stuff like that. It's all there. And I guess if the only thing that we were concerned with at the time was moving my Dad into a facility or trying to find him a place to stay, then I probably would have gotten into that information as thoroughly as possible. But, you know, it went from that to doctors' appointments with my mom, so.

Couples rarely move to assisted living residences together; most residents are on their own [23]. There is little research about transition to assisted living from the perspective of spouses, let alone the perspective of spouses of persons with dementia. Kemp interviewed 20 couples who lived in an assisted living residence together [9], including some where dementia was present. In some couples, both partners had care needs requiring assisted living. For others, the move was because of care needs of one partner, the healthier partner being unable to sustain support at home. These couples had a choice between moving together or one partner moving while the other lived independently. Marital responsibility was a motivator for moving together.

Cost is a factor in the decision to move [7,9,12]. The cost for couples is significant. In the Dementia Transition Study, a man who had dementia described his worries about outliving his savings when he and his wife moved to an assisted living residence. Having lost investment savings, he worried that selling his home would be insufficient *"But that is not going to last that long at 5000 dollars a month, okay. You know. I mean, okay, you'*re *going to hope that you die soon."* Even if just one member of a couple moves to an assisted living residence, the additional cost while continuing to pay household expenses for the family home is a factor in decision-making.

It seems that persons with dementia are unlikely to be meaningfully involved in decision-making about moving to assisted living. In general, decision-making about moving to assisted living, even for people who do not have dementia, seems to be by adult children [12]. Furthermore, while persons with dementia want to be involved in decision-making about their care, they are often excluded, with care partners taking responsibility for decisions [29]. This exclusion is related to care partner concerns about safety and security, care partner stress, the person's lack of insight into problems with functioning and overestimation of their abilities, and feeling pressured by healthcare providers to make decisions without input from the person with dementia [26,29]. Lack of participation in decision-making about the move to assisted living may be explained by a sense of urgency on the part of care partners and ambivalence of persons with dementia.

Larsson et al. [26] reviewed the literature on decision-making about care services for people with dementia, finding that there are three ways that they are involved: They are excluded; their prior preferences are taken into account; or their current preferences are respected. Taking prior preferences into account relies on having had discussions about what the person with dementia wants early in the course of the illness; such discussions are uncommon. Care partners may move from respecting current preferences to convincing the person with dementia to move when they are very concerned about safety. For example, a daughter in the Dementia Transition Study explained

> *She, on the other hand, did not understand or didn't believe … she is accepting she is having a memory issue. She doesn't deny that. She remains to be convinced that it is problematic in safety and sort of ongoing …. So, anyway, we went and saw the doctor. And I had a report for him … and he said 'I think what is in this report is valid and you should sell your house and do that.' But no, she thinks he is part of the conspiracy … So, she is in her house and managing not too badly at the moment. But it's kind of day to day.*

The daughter anticipated that eventually there would be a crisis when she would have to make the decision to move for her mother.

The most commonly found approach to decision-making is exclusion, which includes not telling the person about the decision, telling the person and going ahead with it even if they disagree, and persuading a person over time [26]. Examples of exclusion in decisions about transition of persons with dementia to assisted living include: Two-thirds of the sample in one study moved involuntarily [28], a finding that among people who moved to assisted living, those who were "resistant" to moving were often those with dementia [12], findings that persons with dementia are influenced by recommendations of care partners and healthcare providers [3], and findings that persons with dementia perceive decision-making as being controlled by care partners [24]. Sometimes people with dementia visit the assisted living residence ahead of time or have an overnight stay to test out the residence. This is generally, but not always, experienced positively as involvement in decision-making [25].

Lack of participation in decision-making about moving to an assisted living residence is important because there is some evidence that control over decision-making is associated with satisfaction with the move [9,30] and because of the person with dementia's experience of exclusion. For persons with dementia, being excluded from decision-making about their care threatens identity and is experienced as dehumanizing [26]. A person with dementia in the Dementia Transition Study described being dissatisfied: *"people are making decisions and they are not really consulting me … they have been talking to my wife and she has got a very childish view of the whole thing."* He felt excluded, even though he was able to analyze the reasons for moving, as illustrated in his explanation of accepting but not liking the decision to move:

> *Well, I am trying to be realistic, okay. I am finding that the stairs are a problem, so I can see. Okay, we should be someplace where we don't have to climb stairs but I don't think we have reached a stage where we are incapable of looking after ourselves. But I can see that if I have early onset Alzheimer's, which is what they told me ... then I may reach a stage in the progress of that ailment where I can't look after myself and my wife would not be able to look after me.*

In contrast, another participant in the Dementia Transitions Study described being happy to let his wife choose where they would live, saying *"I let my wife choose. She chose well."*

In summary, although deciding to move to an assisted living residence can happen quickly, it is often a long process of looking for and trying alternate solutions to problems with managing at home independently. Typically, several factors contribute to the decision, including safety of the person with dementia, deteriorating physical health and functioning, caregiver well-being, and access to community-based supports. Ambivalence is common. The extent to which persons with dementia are involved in the decision to move varies. They may be more likely to be excluded from the decision when the move is part of a crisis or as safety concerns mount. Being involved in decision-making may be associated with better adjustment and outcomes.

Moving

Moving involves not only the act of moving but also preparation, including emotional and psychological preparation as well as practical preparation.

Preparing for the move

While the extent to which persons with dementia can prepare depends on how much they are involved in decision-making about moving and how quickly the move comes after a decision is made, when people know about the move in advance, there is a process of anticipation and emotional preparation. Aminzadeh et al. [24] found that persons with dementia anticipated the move "as a major residential change and life transition requiring significant adaptive efforts" (p. 490). Having a positive attitude, accepting the move and "getting on with it" have been described in the literature about older people anticipating moves to assisted living [7]. Similarly, persons with dementia have been described as trying "to focus on their adaptive resources, preserved abilities, and positive personality attributes and coping skills, portraying an image of oneself as a 'survivor'" (p. 492) [24]. Positive anticipation includes having gratitude and a positive outlook about being safe, cared for, and with people [24,25]. It also includes identifying positive attributes that would help the person adjust, including being personable, active, easy-going, and self-confident [24].

While those who identify as extroverts seem more likely to anticipate an easy adjustment, others worry about being lonely or that sensory deficits will make adjustment to the move and communal life difficult [24]. Uncertainty and anxiety are common before the move [22,24,25]. Aminzadeh et al. [24] found that persons with more cognitive deficits were less likely to be able to use cognitive coping strategies to frame the move as positive and that they had more fear of relocation. However, even when persons with dementia have realistic concerns about adjustment, they tend to try to accept the situation, focusing on their coping ability [24]. As a person with dementia explained, *"If you allow it to be, it's rather frightening. But ah, I've faced quite a few challenges in my 90 years. So, I'm not letting this bother me. I think that one just has to assume that everything's going to go well and it probably will go well … I'll survive. I always have"* (p. 492) [24].

Tasks and responsibilities for moving

Tasks that are part of moving include giving up a lease or selling a home; deciding which belongings to take to the new residence; giving away, selling, or disposing of other belongings; arranging for utilities and phone changes or cancellation and change of mailing address; and moving in to the residence. The following quote is a care partner in the Dementia Transitions Study describing preparing for her aunt's move. Her experience illustrates common themes in the literature of being overwhelmed by possessions, the impact of the person with dementia's cognitive impairments on their ability to participate in the process, shared responsibility for assistance by multiple family members, and the emotional challenges of the experience.

> *Well just the running around and kind of organising … it was quite a big apartment and they didn't get rid of anything. So there was a lot of stuff …*

My brother came and got things. So it was to coordinate all of that. Thank God I had my husband to help because he did a lot as well. But you know it was on our mind, because you feel the pressure when you have a deadline …

We hired a company that was recommended by the [assisted living residence]. And, it's one woman looks after seniors, so looks after downsizing and moving. She was actually fantastic, because I had a bit of support in telling yes and no to bring certain things.

Because [person with dementia] could not visualize it. That was the hardest thing, because during the whole month and a half of making decisions as to what to keep and what to give was very frustrating, because of the Alzheimer. Because she would forget what she would say. We literally brought books to the library and I got a call that night asking 'What did you do with my books?' 'We brought them to the library.' 'Well I didn't want to. You pushed me. I wanted to keep those books.' So my husband and I did not sleep. So both of us got up in the morning and said 'You know what? Let's go and get those books back.' And we did.

So there were times where I just wanted to yell at her out of frustration to try and make her understand. But it was difficult. But that woman was able to help and say 'No, you will not be able to fit that.' She went to the apartment made the measurements and was able to say, 'No, you will not be able to fit that in.' … It got to the point near the end, we said 'You know what? Let her bring whatever she needs to bring.' … Some pieces were brought and there was no room and she said 'Well, I can't have them.' And at that point they were thrown out.

Businesses that specialize in assisting older persons with moving can be hired to organize packing, dispersing possessions, and moving. They have not been described in the research about moving to assisted living residences.

Cognitive impairments and problems with instrumental activities of daily living of the person with dementia may become more apparent at this time [22]. For example, a daughter in the Dementia Transitions Study described advocating for her mother, whose landlord was initially reluctant to break the lease:

She had to give her notice to the apartment. And I said 'Are you on a month to month or did you sign a lease?' 'Oh, no, I am on month to month.' So she gave her notice. I just had the insight to speak to the management at the building … They said no, she had just signed a lease in October. So he said 'She will have to sublet or be penalized for 3 months.' I said 'Okay, alright … Can I speak to head office, maybe they will make an exception?' I mean, she has been living there for 29 years. So I phoned and spoke to head office. I said 'Can you make an exception?' She said 'Well, usually they don't care.' … She said 'Wait a second.

Okay, I tore the lease up. She is on a month to month. She is good.' I said 'Oh my God. You are so good. Thank you!'

In the Dementia Transition Study, even when the person with dementia made the decision to sell their home, care partners usually took responsibility for tasks related to moving, sometimes sharing responsibilities among family members. For care partners, this can involve taking time off work and traveling. In some cases, the person with dementia's home may be sold before the move but often the tasks of clearing out possessions and selling the home go on for months after the move. A daughter described taking 2 weeks off work to help with the move. The first week was spent packing and deciding what to take to the new residence. The second week was spent taking things to the new residence and helping her mother settle in. Then, over the following months, she and her siblings tackled clearing out the house while trying to respect her mother's wishes and supporting her settling in to the new residence. Recalling a conversation with her brother, she said

He goes in there and feels so frustrated because he feels overwhelmed. He says, 'I would like to just get a big dump truck and throw everything in there.' But, I mean, it was just something he was saying and [he] wouldn't do that.

Persons with dementia may both appreciate help from care partners and be frustrated or resentful about not being in control of the process [24]. Mixed reactions to disbursement of belongings have been reported; sadness about losing objects with sentimental value, relief at not being burdened by belongings that had become too much to manage, and being physically and emotionally overwhelmed by the process [24]. For care partners, the physical work and their emotions can be challenging. In the Dementia Transitions Study, care partners described being overwhelmed and frustrated. A daughter who had a chronic illness spoke about the work being tiring and about the meaning of losing her childhood home:

Like I cried last night. I was in a lot of pain, and ... I was actually in an extreme amount of pain and it could have been partially from all that I had done. But, it was the turn of events. I mean this is it. I grew up in that house. I mean I was 10 when I moved into that house.

Settling in

Two overlapping phases of adjustment after the move are described in the literature. The first is settling in and initial adjustment in the first several weeks. The second is recreating home, when the person comes to feel at home in the new residence, a process that can take several months and may never be completely achieved.

Aminzadeh et al. [22] found that the days and weeks after the move were an intense, difficult transition period for most persons with dementia. This initial adjustment was described as shocking, disruptive, and a time of confusion, disorientation, and distress. The challenge and negative experiences were greater for those with worse cognitive functioning. These negative experiences included: deteriorating cognitive functioning; anxiety symptoms, including panic; depression symptoms, including passive suicidal thoughts; loneliness; and, suspiciousness. For some, there was a worsening of comorbid physical illness, with three of 16 people in the study requiring hospital admission. A care partner described this process:

> *At first she felt a bit awkward. When we dropped her off she was very anxious ... She didn't know where anything was ... She has lost some time perspective ... It started to cause her anxiety and blood pressure went up. So, we were shuttling her back and forth between doctors ... There was quite a bit of commotion. ... She was uncomfortable, just the whole thing ... there was a huge disruption ... She started doing what they called "hoarding" (p. 227).*

Eventually, the person with dementia adjusts and settles in to the new home and new routine. Thein et al. [25] found that by about 9 weeks after the move, most persons with dementia felt settled and happy, while continuing to miss their independence. However, persons with more severe cognitive impairment may have recurring periods of feeling unsettled and lost. As described by a care partner, *"It goes in cycles and I don't know what triggers it. All of a sudden she'll start phoning ... not knowing what to do, where things are ... feeling lost, missing her previous arrangement."* (p. 228) [22]. In the Dementia Transitions Study, it was common for persons with dementia to continue to talk about home, want to go home, and try to go home—but with decreasing frequency over time.

Challenges for the person with dementia in settling in include finding their way around, feeling confined in the small space of their room or unit, missing their kitchen and preferred foods, missing being engaged in housekeeping and home maintenance activities, missing their neighborhood, having less privacy, and learning and adjusting to the policies, rules, and practices of the assisted living residence. These rules and practices (e.g., meal schedules, dining routines, requirements for residents and visitors to sign in and out of the building) are described as restrictive and intrusive

[22,31]. A person with dementia described restrictions to autonomy and meaningful activities and her desire to be outside: *"But I'm not* allowed *to feed the birds here ... but they don't let you just sit out there by yourself; one of* them [staff members] *has to be out there with you"* (p. 132) [31].

Another challenge is beginning to feel connected and engaged while adjusting to living with frail older persons [22]. Some persons with dementia dislike or resent being with old, frail people with cognitive impairment, at least initially [22,31]. Others feel a sense of relief and acceptance being with older people who have similar disabilities or frailty who are *"all in the same boat"* (p. 230) [22]. This is consistent with research findings that residing with peers who have dementia can reduce feelings of loneliness and isolation and increase feelings of acceptance [32].

After moving to an assisted living residence, persons with dementia may feel lonely and disconnected, missing friends and neighbors, which may be partly attributable to stigma and friends' avoidance of the assisted living residence [22]. A care partner described

> *I think she must feel a bit lonely ... [be]cause I noticed her friends don't drop in to see her ... the other reason is that, you've to sign in. I think there's a sense that this is a more protected environment" (p. 228) [22].*

It is difficult to form new connections. Fitting in at meal times, the most important social event of the day in an assisted living residence, can be difficult, as described by this person with dementia:

> *Every time I went to sit at a table I was told, "Oh, so-and-so sits there." It was quite difficult for a while ... until I met this friend and he said, "Come sit here." Then I had a place where I felt I belonged ... That sounds a very small thing, but believe me, it's a very important thing to feel that you're welcomed somewhere (p. 229) [22].*

Sensory deficits and cognitive impairment can make fitting in more difficult, as described by this person with dementia who had a hearing impairment:

> *I simply cannot follow the conversation ... It's just plain frustrating to sit there and see them talking and you have no ideas what you are doing there ... Your train of thought is disturbed and you want to say something and you lost it (p. 230) [22].*

Staff in the assisted living residence play an important role in settling in. According to care partners, having friendly conversations with staff is important to relieve new residents' loneliness [22,25]. Furthermore, directions and introductions from staff help the person get to know the routine of life in the assisted living residence, including activities and mealtime rituals and practices [22]. A care partner explained the importance of the positive atmosphere created by staff,

> *The staff are here because they want to be; they seem to really enjoy the work, and they're so good to the people; they're just like family. And they sit down and eat with them, so they are like family (p. 46) [33].*

Some research indicates that it may be more difficult for families to receive information they can use in the transition; that policy clarity significantly influences the process [33]. Lack of clear communication, coupled with the move being a stressful and emotional experience for care partners, can make it difficult for them to take in information they receive, leaving them open to confusion and frustration [33]. It is important that families can access information in different ways before and after the move. They need information about what care is provided and what requires additional fees, how to communicate with staff, what to expect in terms of communication from staff and managers, how health care and medical emergencies are handled, and when activities are scheduled [33]. In the Dementia Transitions Study, some care partners were not sure what level of care was being provided. This was particularly challenging when the person was receiving home care through publicly provided home care services in addition to the basic support in the assisted living residence. Others described the importance to them of knowing that the staff were keeping the person with dementia safe, appreciating that staff were watching out for the residents. For example, a daughter described,

> *I don't know whether she was on her way out the door or she stopped at something at the desk, but she was on her way home. She was going to walk home. And so, they phoned my brother and he talked to Mom. He's very calm. He said 'Mom, you need to go back to your apartment and I'll be there shortly.' And so she did. She went back and they had a very good talk to her.*

The extent to which staff in assisted living residences are knowledgeable about dementia varies. Care partners worry when they observe that staff do not understand the support needs of persons with dementia [33]. Care partners in the Dementia Transitions Study talked about needing to advocate for the person with dementia:

> *But I have to keep track of things going on over there because ... I mean, little things. Privacy issues that my mom had had. I went to the director of care and*

> *told him. I mean you have to be involved in these places because if you're not it gets out of hand.*

Care partners are described as "vital" in supporting the process of settling in (p. 14) [25]. Consistent with research about the transition to an assisted living residence, care partners in the Dementia Transitions Study expected that it would take at least 2 months for the person with dementia to adjust to their new home and they expected that there would be a period of stress for themselves before they would feel relieved.

In the initial few weeks, most care partners were calling every day and visiting several times a week. They would either call themselves or talk when the person with dementia called for reassurance, for help when they had lost something, or because they wanted to go home. A care partner described, *"She'll call me up in the middle of the day, just wanting to chat for a minute. You know, hear your voice. That's fine with me. I'm quite happy to do that."* Another care partner described more frequent calls and being upset about not being able to respond to them all. Care partners also supported adjustment by hiring or arranging extra help for the first little while, helping the person continue to attend community day programs, helping them understand that they are in their new home, and helping them to feel comfortable with the services in the residence. For example, a care partner described:

> *She said to me, 'I always ask for the* smaller *portion but they don't listen to me.' And I said, 'That's okay, just eat what you can and just leave the rest. They will not be offended by it. If you are not capable of eating them all, don't eat them all.'*

Similar to other research [22], after the move, care partners in the Dementia Transitions Study continued to support health and well-being of the persons with dementia by attending healthcare appointments with them, encouraging physical activities, reminding them to attend programs, and refilling prescriptions. They ran errands. Some did laundry and changed bed linens.

Recreating home

While most people with dementia eventually settle in and get past the difficult initial transition period, not everyone comes to feel at home. The extent to which this happens depends on many factors including policies and practices in the assisted living residence, environmental supports for the person with dementia, support from the care partner, the person's premorbid personality and coping style, and the person's dementia symptoms and cognitive abilities [22,33].

A significant aspect of feeling at home and of quality of life for persons with dementia is feeling connected to place and having relationships within and outside the home [4]. Furthermore, connectedness is an important aspect of quality of life in assisted living residences [34]. Connectedness and relationships are disrupted in the move to an assisted living residence. By 2–6 months after the move, some persons with dementia are engaged in activities and have made new friends [22,33]. For some, this may be easier in residences that specialize in dementia care or where there are more people with dementia, as explained by a person with dementia: *"I like the people that have the same disease I have, so it makes it intimate with other people because people that don't have this disease don't understand it. But they do here."* (p. 46) [33]. Cognitive impairment that affects the person's social communication and behavior may lead to isolation from residents who are offended by or intolerant of such behavior [22].

Aminzadeh et al. [22] found variability in comfort with people and activities in the residence 2–6 months after the move and that many people did not "fit in" socially. There were three types of experiences fitting in to social life and congregate living. First were those who saw themselves as extroverts and had confidence to engage in activities and talk to new people. Second were those who felt lonely and spent their time alone. Care partners described them "'cocooning' in their rooms" (p. 232) where they connected with memories of past relationships through their belongings. Third, were those who were described as "being on the fringe" (p. 232), who did not join activities but spent solitary time in public areas or in passive group activities, engaging in superficial interactions with other residents.

A significant part of adjustment is maintaining connections with family and with family traditions [33]. This includes regularly attending family gatherings such as meals and celebrations. When it is not possible for the person with dementia to leave the residence, family may gather there instead. Photographs are used to share information about events the person cannot attend. Visitors may view more distant relatives' social media images and posts with the person with dementia. Telephone calls are used to keep in touch with family. Stadnyk et al. [33] found that characteristics of the assisted living residence supported maintaining family connections, including staff assisting residents to prepare for family outings, having space in the resident's private living area to display photographs, and access to telephone, including support to use the phone if needed.

Activities in the assisted living residence are a way to feel connected and engaged. Depending on the activities that the person enjoys, having private space may allow the person with dementia to continue past activities, such as watching television or reading. However, some previously enjoyed and meaningful activities may not be feasible in the assisted living residence and the person with dementia may miss these activities, even if they were unable to do them before they moved. Organized activities in the assisted living residence allow the person with dementia to develop new interests, sometimes to the surprise of care partners [33].

Engaging in meaningful activities in the residence can build attachments to people and the residence and are important for quality of life of persons with dementia [2,4,30]. However, apathy, which is common in persons with dementia, may inhibit residents from participating in activities which are themselves an antidote to apathy and diminished quality of life [35]. Care partners can remind residents with dementia to participate in activities, and thus, it is important that the assisted living residence keeps families updated about activities [33]. Care partners have reported that it is important that staff accommodate for dementia symptoms that may inhibit participation. *"I think that my mom needs that cuing now, and I don't think that [staff on the unit] provides it, but the regular activity staff, they provide it, and then she does go."* (p. 47) [33].

Being outdoors and in nature contributes to quality of life and is an important aspect of the meaning of home for persons with dementia [2,4]. As previously described, a challenge of the move to an assisted living residence is that it is usually a move away from a familiar neighborhood, where it felt safe to walk. People with dementia are often not able to get to know and walk in their new neighborhoods— because of either restrictions on exiting the residence or their diminished abilities to navigate and find their way around. Assisted living residences vary in the extent to which outdoor spaces are fully accessible to those with dementia [33]. Consistent with best practices in environmental design for persons with dementia [36], care partner participants in research about relocation to assisted living have noted the importance that organized activities include the opportunity to be outside and to go for walks [33].

Other aspects of the environment that contribute to a quality living environment for people with dementia include home-like private rooms and common areas, having personal objects in private areas, high levels of ambient and natural light, food quality, home-like dining experiences, clear sight lines, design that supports orientation, including art that reflects the seasons and local scenery, readable calendars and clocks, views of the outdoors and nature, readable signage, and ease of wayfinding [22,33,36]. Incorporating these aspects of design in assisted living residences supports adjustment [34].

Aminzadeh et al. [22] found that over time, persons with dementia engaged in an internal process of finding meaning in the transition, which was symbolic of a longer transition from seeing oneself as independent, competent and autonomous to being dependent, less engaged, surrounded by disability and aging. Some seemed to find it helpful to frame the move as a natural consequence of aging, as illustrated by this person with dementia:

> *When you get older, I think it's a good idea to go into a place like this. … I think that if the weather's bad, it's freezing rain coming down, it doesn't bother me at all. I can just stay right here and watch it happen. … There is a time in your life when you're going to have to change your lifestyle and change your friends, even though you wish you didn't have to. (p. 233).*

A minority of participants in this study continued to grieve the loss of home. They focused on the negative aspects of the move, unable to recall the challenges they experienced in their former homes. They maintained hope by viewing the move as temporary. Similar to findings of Thein et al. [25], by 6 months after the move, most participants in this study [22] fell between these two extremes, accepting the move, having recreated some aspects of home at the assisted living residence, while still thinking about and missing their former homes.

Care partners may not be aware of the challenges that persons with dementia experience with psychological adaptation during this transition [22], suggesting that interventions to support the process may be helpful. Life review or reminiscence may be appropriate to support the transition to living in an assisted living residence for persons in earlier stages of dementia. These approaches are widely used, but there is limited evidence of effectiveness [37,38]. O'Hora et al. [39,40] examined a family life review intervention with 14 people who moved to an assisted living residence and their relatives. The sample included some persons with mild or moderate dementia. Overall, the intervention was acceptable to participants. It seemed to be more meaningful for relatives than residents. It was not suitable for families experiencing significant turmoil, and it was challenging for persons with substantial memory problems [40].

As the person with dementia transitions from settling in to recreating home, care partners continue to provide emotional and instrumental support. They continue to advocate with staff and managers in the residence. They also attend residence activities with the person with dementia. Healthcare providers may assume that by the time care partners get to this transition they have a good understanding of dementia and how it affects functioning, but this is not necessarily true. Aminzadeh et al. [22] found that some care partners were unable to understand how dementia affected the person's ability to engage in activities and integrate into life at the residence. This indicates the importance of providing families with information and education about the adjustment process in relation to symptoms and cognitive impairments associated with dementia.

In the Dementia Transitions Study, while transitioning to the assisted living residence, care partners were already thinking about the next transition. Policies and practices related to discharge from assisted living residences and transfer to long-term care homes vary [41,42]. When an assisted living residence has a dementia care unit, the person with dementia may be able to move to it if space is available and the person has the means to pay for it. Mead et al. [42] found that care partners who advocated for their relatives with dementia and worked with residence managers to decrease problematic behaviors were able to prevent discharge to long-term care.

Care partners in the Dementia Transitions Study expected that the person with dementia would eventually deteriorate, need more care than available in the residence, and have to move to a long-term care home. Two of them hoped that the person with dementia would die before that transition came to pass *"I mean to be brutally frank, I hope he dies before he has to move anywhere else. I think he does too. Because it's not going to be easy to make that decision."* Some saw the possibility of paying for extra services as a way of avoiding or putting off another transition. Others did not have funds to pay for more services, as illustrated by a care partner in the Dementia Transitions Study, *"She's okay but, I mean money is getting. We are all kind of a little bit nervous and we have applied for long-term care because the place where she is at is very expensive."* Another care partner talked about not being able to afford services beyond the base rate,

> *But you have to pay extra for it, and we can't afford it. We just don't have that kind of money. A lot of people here are military. So they have good pensions and whatever extra job they had. So they are well provided for. We don't have that luxury. If he needs extra care, we can't afford it. So I don't know what's going to happen.*

Practices and policies that support the transition to an assisted living residence

Clearly, the transition from home in the community to home in an assisted living residence is meaningful and challenging. First, a caveat. We identified a few studies that focused on the transition to an assisted living residence for persons with dementia, so there was limited replication of findings. There is a need for more research, particularly longitudinal research with representative samples. Research is needed about the experience of this transition in subpopulations of persons with dementia, including ethnocultural and gender-diverse minority groups. Selection bias is part of existing research with samples of people who have made the transition. This means that we do not know enough about how decisions to move are made over time or about outcomes for people who decide against or cannot afford this transition.

Accessibility of assisted living residences is an issue. This is important given the trend for assisted living to be an alternative to long-term care homes, shifting responsibility for care for persons with dementia from the state to the individual. Choices regarding assisted living are highly constrained by financial resources and the option of assisted living is only available to those who can afford it. The second factor that may affect accessibility is the vital role that care partners play in this transition; those who do not have care partners may not be able to navigate it.

Healthcare providers and care partners should not assume that just because a person has dementia they cannot be involved in decisions about assisted living. Interventions aimed at improving shared decision-making developed for care partners and for care partner-person with dementia dyads may be helpful in the context of discussions about moving to an assisted living residence [29]. Timely planning for the possibility of moving to an assisted living residence is recommended, to give the person with dementia time to accept and prepare for the transition [3,30].

Having access to good information is important for decision-making about whether to move and where to move, as well as for adjustment after the move. To make an informed choice about where to move, people need to know the costs, how costs will increase if additional care is needed as dementia progresses, and the discharge policies of residences they are considering. It is important that family members and care partners are provided with information about the assisted living residence before and after the move, including information about the scope of support and care provided by the assisted living residence, their responsibilities, who to contact with questions and concerns, and when and where activities occur.

Care partners' coping during the transition may be challenged. Furthermore, they may not understand how the person with dementia is affected by it. Some care partners may benefit from psychoeducational support interventions, like those that are used to support care partners around the move to a long-term care home [43]. Gaugler et al. [44] pilot-tested a six-session intervention with a transition counselor for care partners of persons with dementia who moved to an assisted living residence or a long-term care home. The intervention was promising, with potential beneficial effects on caregiver stress and overload. As discussed, family life review may be helpful for some care partners but may not be feasible for persons with dementia [39].

Characteristics of the assisted living residence, including design, policies and practices, effect adjustment, and quality of life after the move. Most assisted living residences have mixed populations of residents with and without dementia, and most are not designed for dementia care. If dementia-friendly design elements are present, the transition to recreating home in the residence should be easier for persons with dementia. It is important that managers and staff in assisted living residences understand dementia and are able to adapt practices to meet the needs of residents who have dementia and their care partners. For example, a recommended practice in supporting new residents of assisted living residences is promoting social activities and engagement [30]. Adapting this practice for new residents who have dementia, staff may need to repeatedly remind people about activities and be more active in encouraging participation.

The following practices were identified as supporting adjustment after a move to a long-term care home or assisted living residence [30] and may benefit persons with dementia:

- Expect and acknowledge loss and grief, providing ongoing opportunity for discussion of emotions and adjustment;
- Have a volunteer or staff member regularly meet with the new resident;
- Promote furnishing the resident's room with belongings from home;
- Help residents get to know and connect with other residents, staff, and volunteers;
- Preserve personal space and privacy;
- Allow residents to participate in domestic tasks;
- Promote opportunities for residents to help one another;
- Help residents and family members stay in touch;
- Provide opportunities for personally meaningful activities appropriate to residents' abilities.

Conclusion

Adjusting to a move is a challenge for anyone. Moving out of one's home in the community into an assisted living residence is highly meaningful, and the process of adjusting to that move is challenging. Dementia makes that adjustment yet more challenging.

We conclude with a return to the meaning of home for persons with dementia. A systematic meta-synthesis identified four elements of a person with dementia's experience of lived space: belonging, meaningfulness, safety and security, and autonomy [2]. Belonging in one's home is being with familiar objects, activities, and people. It connects people with dementia to their history, it is more than the house or apartment, and it includes the neighborhood too. Meaningfulness refers to meaning that comes from activities, connections to other people, socializing, and having a private space for solitude and rejuvenation. Being able to be outdoors at home brings meaning. Recreating a sense of belonging in a new space takes time and effort on the part of the person with dementia, their care partners, family, and staff in the assisted living residence. The extent to which belonging and meaningfulness are part of life after the move is influenced by factors within the person with dementia, including cognitive abilities, coping style, personality, comorbid conditions, as well as financial resources. It is also influenced by factors in the social and physical environment, including support from family and friends, policies and practices in the assisted living residence, and design of the assisted living residence.

The third element of the lived space of persons with dementia is safety and security. Safety and security at home diminishes as dementia advances and is a major reason for the move to an assisted living residence. The feeling of safety and security after the move contributes to accepting loss of the previous home. Concerns about safety and security motivate care partners to consider assisted living and are a reason for taking over decision-making.

The fourth element of the lived space of persons with dementia is home as a place of autonomy, control, and choice. Having autonomy, choice, and control is central to quality of life in dementia [4]. Autonomy at home decreases as dementia progresses. Many persons with dementia do not have choice about whether or not to move to an assisted living residence. The model of assisted living residences was developed with the aim of supporting resident autonomy and choice, as an alternative to institution-based care of long-term care homes. However, in practice, autonomy and choice are constrained, especially for people with dementia. There is a tension between the original philosophy of assisted living and the reality that assisted living residences are increasingly providing high levels of care to persons with dementia, foregoing autonomy in favor of safety. While the idea of autonomy associated with assisted living homes is appealing, it may be more an idea than reality for persons with dementia and must be balanced against possibility of worse experiences for them in assisted living residences than in long-term care homes. Long-term care homes are more regulated than assisted living residences are, and it has been suggested that current assisted living regulations insufficiently promote quality care for persons with dementia [14]. We do not have good evidence about whether assisted living residences, or long-term care homes provide better outcomes for people with dementia.

Relationships and sense of place are core elements of quality of life of persons with dementia [4]. The evidence in this chapter is that for the most part, persons with dementia who move to assisted living residences are resilient and can sustain or regain meaningful relationships and sense of place. However, many only partly achieve this. We need more research about ways to make it possible for people experiencing this transition to having meaningful relationships and attachments to and in their new homes.

References

[1] Molony SL. The meaning of home. Res Gerontol Nurs 2010;3(4):291–307.
[2] Førsund LH, Grov EK, Helvik AS, Juvet LK, Skovdahl K, Eriksen S. The experience of lived space in persons with dementia: a systematic meta-synthesis. BMC Geriatr 2018;18(1):1–27.
[3] Aminzadeh F, Dalziel W, Molnar F, Garcia L. Meanings, functions, and experiences of living at home for individuals with dementia at the critical point of relocation. J Gerontol Nurs 2010;36(6):28–35.
[4] O'Rourke HM, Duggleby W, Fraser KD, Jerke L. Factors that affect quality of life from the perspective of people with dementia: a metasynthesis. J Am Geriatr Soc 2015;63(1):24–38.
[5] Drummond N, McCleary L, Garcia L, McGilton K, Molnar F, Dalziel W, et al. Assessing determinants of perceived quality in transitions for people with dementia: a prospective observational study. Can Geriatr J 2019;22(1):13–22.
[6] Fonad E, Wahlin TBR, Heikkila K, Emami A. Moving to and living in a retirement home. J Hous Elder 2006;20(3):45–60.
[7] Chen S, Brown JW, Mefford LC, de La Roche A, McLain AM, Haun MW, et al. Elders' decisions to enter assisted living facilities: a grounded theory study. J Hous Elder 2008;22(1–2):86–103.
[8] Rockwell J. Settling in or just settling? Exploring older adults' narratives of relocation to assisted living. UBC; 2017.

[9] Kemp C. Negotiating transitions in later life: married couples in assisted living. J Appl Gerontol 2008;27(3):231–51.
[10] Kisling-Rundgren A, Paul DP, Coustasse A. Costs, staffing, and services of assisted living in the United States: a literature review. Health Care Manag 2016;35(2):156–63.
[11] Fields NL, Koenig T, Dabelko-Schoeny H. Resident transitions to assisted living: a role for social workers. Health Soc Work 2012;37(3):147–54.
[12] Ball MM, Perkins MM, Hollingsworth C, Whittington FJ, King SV. Pathways to assisted living: the influence of race and class. J Appl Gerontol 2008;28(1):81–108.
[13] Garner R, Tanuseputro P, Manuel DG, Sanmartin C. Transitions to long-term and residential care among older Canadians. Health Rep 2018;29(5):13–23.
[14] Kaskie BP, Nattinger M, Potter A. Policies to protect persons with dementia in assisted living: deja vu all over again? Gerontol 2015;55(2):199–209.
[15] Rosenblatt A, Samus QM, Steele CD, Baker AS, Harper MG, Brandt J, et al. The Maryland Assisted Living Study: prevalence, recognition, and treatment of dementia and other psychiatric disorders in the assisted living population of central Maryland. J Am Geriatr Soc 2004;52(10):1618–25.
[16] Zimmerman BS, Sloane PD, Reed D. Dementia prevalence and care in assisted living. Health Aff 2014;4(4):658–66.
[17] Carder PC. State regulatory approaches for dementia care in residential care and assisted living. Gerontol 2017;57(4):776–86.
[18] Fields NL, Richardson VE, Schuman D. Marital status and persons with dementia in assisted living: an exploration of length of stay. Am J Alzheimer's Dis Other Dementias 2017;32(2):82–9.
[19] Peeples AD, Frankowski AC, Roth EG, Eckert JK, Morgan LA, Nemec M, et al. The facade of stability in assisted living. Journals Gerontol Ser B 2014;69(3):431–41.
[20] McCleary L, Persaud M, Hum S, Pimlott NJ, Cohen CA, Koehn S, et al. Pathways to dementia diagnosis among South Asian Canadians. Dementia 2012;12(6):769–89.
[21] Cole L, Samsi K, Manthorpe J. Is there an "optimal time" to move to a care home for a person with dementia? A systematic review of the literature. Int Psychogeriatr 2018;30(11):1649–70.
[22] Aminzadeh F, Molnar FJ, Dalziel WB, Garcia LJ. An exploration of adjustment needs and efforts of persons with dementia after relocation to a residential care facility. J Hous Elder 2013;27(1–2):221–40.
[23] Kemp CL. Married couples in assisted living: adult children's experiences providing support. J Fam Issues 2011;33(5):639–61.
[24] Aminzadeh F, Dalziel WB, Molnar FJ, Garcia LJ. Symbolic meaning of relocation to a residential care facility for persons with dementia. Aging Ment Health 2009;13(3):487–96.
[25] Thein NW, D'Souza G, Sheehan B. Expectations and experience of moving to a care home: perceptions of older people with dementia. Dementia 2011;10(1):7–18.
[26] Larsson AT, Österholm JH. How are decisions on care services for people with dementia made and experienced? A systematic review and qualitative synthesis of recent empirical findings. Int Psychogeriatr 2014;26(11):1849–62.
[27] Poyner C, Innes A, Dekker F. Extra care: viable for couples living with dementia? Hous Care Support 2017;20(1):8–18.
[28] Kelsey SG, Laditka SB, Laditka JN. Caregiver perspectives on transitions to assisted living and memory care. Am J Alzheimers Dis Other Demen 2010;25(3):255–64.
[29] Miller LM, Whitlatch CJ, Lyons KS. Shared decision-making in dementia: a review of patient and family carer involvement. Dementia 2016;15(5):1141–57.

[30] Brownie S, Horstmanshof L, Garbutt R. Factors that impact residents' transition and psychological adjustment to long-term aged care: a systematic literature review. Int J Nurs Stud December 2014;51(12):1654–66.
[31] Jaffe DJ, Wellin C. June's troubled transition: adjustment to residential care for older adults with dementia. Care Manag J 2008;9(3):128–37.
[32] Eriksen S, Helvik AS, Juvet LK, Skovdahl K, Førsund LH, Grov EK. The experience of relations in persons with dementia: a systematic meta-synthesis. Dement Geriatr Cognit Disord 2016:342–68.
[33] Stadnyk RL, Jurczak SC, Johnson V, Augustine H, Sampson RD. Effects of the physical and social environment on resident-family member activities in assisted living facilities for persons with dementia. Seniors Hous Care J 2013;21(1):36–52.
[34] Bradshaw SA, Playford ED, Riazi A. Living well in care homes: a systematic review of qualitative studies. Age Ageing 2012;41(4):429–40.
[35] Brodaty H, Burns K. Nonpharmacological management of apathy in dementia: a systematic review. Am J Geriatr Psychiatry 2012;20(7):549–64.
[36] Waller S, Masterson A, Evans SC. The development of environmental assessment tools to support the creation of dementia friendly care environments: innovative practice. Dementia 2017;16(2):226–32.
[37] Woods B, Spector AE, Jones CA, Orrell M, Davies SP. Reminiscence therapy for dementia (Review). Cochrane Database Syst Rev 2005;(2):1–36.
[38] Lawrence V, Fossey J, Ballard C, Moniz-Cook E, Murray J. Improving quality of life for people with dementia in care homes: making psychosocial interventions work. Br J Psychiatry 2012;201(5):344–51.
[39] O'Hora KA, Roberto KA. Navigating emotions and relationship dynamics: family life review as a clinical tool for older adults during a relocation transition into an assisted living facility. Aging Ment Health 2019;23(4):404–10.
[40] O'Hora KA, Roberto KA. Facilitating family life review during a relocation to assisted living: exploring contextual impact on family adjustment. Clin Gerontol 2019;42(3):323–33.
[41] Golant SM. Do impaired older persons with health care needs occupy U.S. assisted living facilities? An analysis of six national studies. J Gerontol Ser B 2004;59(2):S68–79.
[42] Mead LC, Eckert JK, Schumacher JG, Zimmerman S. Sociocultural aspects of transitions from assisted living for residents with dementia. Gerontol 2005;45(Suppl. 1):115–23.
[43] Müller C, Lautenschläger S, Meyer G, Stephan A. Interventions to support people with dementia and their caregivers during the transition from home care to nursing home care: a systematic review. Int J Nurs Stud 2017;71:139–52.
[44] Gaugler JE, Reese M, Sauld J. A pilot evaluation of psychosocial support for family caregivers of relatives with dementia in long-term care: the residential care transition module. Res Gerontol Nurs 2015;8(4):161–72.

CHAPTER 8

Relocation to a long-term care home

Annie Robitaille
Interdisciplinary School of Health Sciences, University of Ottawa, Ottawa, Ontario, Canada

Introduction

As the population is increasing rapidly in age, the number of people living with dementia is also increasing and is expected to triple by 2050 [1–4], making dementia a global challenge requiring worldwide attention. As clearly stressed by the Lancet Commission on Dementia Prevention, Intervention, and Care, although a cure for dementia has not yet been found, several actions can be taken to considerably improve the lives of those affected by dementia [5].

As demonstrated by the previous chapters, in addition to the everyday changes that people living with dementia go through (e.g., memory loss, difficulty performing tasks, changes in physical abilities, mood, and behaviors), these individuals experience many transitions in a relatively short period (e.g., initial diagnosis of dementia, loss of financial independence, driving cessation, and change in living arrangement) [6]. These transitions are an extreme source of stress for all those involved. To improve the dementia journey of those affected by dementia, it is imperative that we better understand these transitions so that initiatives be put in place that are tailored to their needs. The previous chapters have discussed many of the transitions lived by people with dementia. This chapter will focus specifically on the transition of moving from one's home to a long-term care facility (throughout this chapter, I use the term long-term care home instead of long-term care facility).

In this chapter, I use the term long-term care home to describe a residential facility that provides a variety of services (e.g., living accommodation, medical and personal care, assistance with everyday activities) and 24 h, 7 days a week supervision that is not hospital-based. Other terms used to describe long-term care homes across different countries includes nursing home, care home, hospice home, skilled-nursing facility, assisted-living facility, residential continuing care, personal care facility, and aged

Evidence-informed Approaches for Managing Dementia Transitions
ISBN 978-0-12-817566-8
https://doi.org/10.1016/B978-0-12-817566-8.00008-5

care [7]. Long-term care homes provide around the clock care to individuals with complex health needs over an extended period.

The accumulation of stressful events and transitions and the lack of time to recover from one transition before the next one occurs makes this, often final, transition that much more difficult to adjust to. The potential impact of this transition is further illustrated by the fact that it has been approved as a diagnosis called Relocation Stress Syndrome (i.e., includes depression, anxiety, apprehension, and loneliness) by the North American Nursing Diagnosis Association [8,9].

With the aging of the population and the increased number of people living with dementia, the number of people who are dependent on others to carry out basic tasks of daily living is increasing. Fortunately, this has led to an increased number of initiatives to help people with dementia age in place [10]. However, the aging in place strategy is not possible for everyone, making the transition to a long-term care home inevitable for many. Not only are these individuals having to leave their home (many of these individuals have lived in their home for the majority of their lives), but they are also having to leave behind the social environment that supported them. That is, family members might be less likely to visit, neighborhood friends are lost, and activities around the house are no longer possible. Understanding the factors that ease or aggravate this transition is key to improving the quality of life of people living in long-term care.

Those affected by the transition include not only the people with dementia but also care providers and loved ones [11]. For example, before the move, caregivers of people with dementia need to contemplate the right moment for the move and find a long-term care home deemed appropriate for the needs of their loved one. After their loved ones have moved into long-term care, caregivers continue to be at an increased risk of depression and physical ailments as their sense of direct and indirect responsibility continues despite their loved one no longer living in their home [11–15]. For exhausted caregivers, who have seen this transition as a last resort, it is hard for them to keep a sense of "home" as their loved one now lives in a home that is regulated by the state and, in some cases, might focus more on safety and care, than quality of life [16,17]. For caregivers to feel a sense of relief, they need to feel, understand, and trust that their loved one is socially engaged and has the best quality of life possible.

The first purpose of this chapter is to review the current status of research on the transition of moving to a long-term care home for people living with dementia and their caregivers. Throughout the chapter, qualitative results (i.e., quotes from people with dementia and their caregivers) taken from the Dementia Transition study, a Canadian, 24-month, prospective longitudinal study that examined the transitions lived by people with dementia over time, are presented (Chapter 1). For more detailed information about this study, refer to the study by Drummond et al. [18]. These quotes further illustrate information expressed throughout the chapter and emphasize the importance of people's perceptions when studying this transition. The second purpose of this chapter is to highlight the importance of longitudinal studies when studying this transition, so that the experiences and adaptation of people with dementia and their caregivers can be better understood. Implications for informing and influencing important decision-making processes about the transition to long-term care and areas needing more research are addressed.

What is the transition from home care to long-term care?

As with many of the transitions mentioned in the previous chapters, this transition is not necessarily characterized by an abrupt change [11]. Rather, this transition often includes a period of time before and after the actual move to long-term care and the length of this transition varies extensively from one individual to the next and between the person with dementia and caregivers. For example, some individuals might contemplate the move for months before actually starting the process and may be on a waiting list for 2 years before actually moving to a long-term care home, whereas others might need to move within a few weeks because of a change in functioning. The decisional process that caregivers must go through was illustrated by caregivers who participated in the Dementia Transition study:

I know eventually, I'm afraid, he will get to the point where he is going to have to go into a home. When that point comes we will cross that bridge when we get there.

I'm struggling right now with it, because I don't want to screw up the process, but at the same time, is he going to need it in a year, is he going to need it in two years, is he not going to — I don't know, but as long as he has the option of rejecting an opportunity.

Once in their new setting, the transition continues, during which the person with dementia and their caregiver must adjust to this change. Again, this varies from one individual to the next with some individuals adjusting more quickly to their new environment than others.

The conceptualization of this transition is important as it has implications for researchers (e.g., it highlights the need for more longitudinal research examining the trajectories of people with dementia and their caregivers before and after transitioning to long-term care) and for policy change and program development. To a great extent, the approach by which long-term care systems are run is systemic rather than being focused on the individual. As demonstrated in this chapter, individual differences in how people with dementia and their caregivers go through the transition from home care to a long-term care home highlight the imminent need for more individualized support during this transition.

Factors that precipitate the transition to long-term care

A number of factors have been identified that precipitate or delay transitioning to long-term care [19–24]. A recent scoping review by Merla et al. identified a number of factors related to an individual transitioning from home care to long-term care from the perspective of caregivers [19]. Factors that were identified as precipitating the move included deterioration in the psychological and physical health of people living with dementia (e.g., stroke, fractures, incontinence, vision loss, increased comorbidities) and/or the caregiver (e.g., depression, stress, cancer diagnosis), an increase in behaviors related to dementia (e.g., physical aggression, wandering), accidents occurring in the house which jeopardize the safety of the person living with dementia or the caregiver (e.g., fire, falls), interference with employment responsibilities and social life, lack of support from other family members, friends, and healthcare professionals, receiving approval from others that it is acceptable to transition to long-term care (i.e., this reassurance reduces the sense of guilt associated with this transition), and the emotional state of caregivers before the move (e.g., ambivalence, powerlessness, worry). These factors were also mentioned by caregivers in the Dementia Transitions Study. When asked about the reason for considering relocation to a long-term care home, caregivers mentioned

It was day to day requirements. You know we got a problem with aggression and we just didn't want to take a chance.

We didn't want her to fall and we were scared of her taking the stairs and that. I would love in my heart that we had all the facilities at home to be able to take care of her but in this present day there isn't.

He was totally needing help that I would have had to hire someone permanently to help me day and night.

Caregivers also mentioned their own health decline as a reason for considering relocating their loved to a long-term care home. For example, *"He fell, I couldn't lift him up, so I would drag him off. I was having more and more blackouts."*

On the other hand, caregivers also identified factors such as family conflict (e.g., some family members are against placing their loved one in a long-term care), cultural expectations against long-term care and lack of cultural sensitivity of the home, sense of duty with regards to caring for the person, negative perception of long-term care facilities in general, and lack of suitability of the home as factors delaying the transition to long-term care [19]. Being aware of these factors can help to determine when individuals living with dementia are most likely to move to a long-term care home and help to better prepare and meet their needs during the transition. In addition to understanding the predictors of moving to a long-term care home, it is also important to better understand the factors that can improve the transition for people with dementia and their caregivers.

Factors that influence the experience of a person with dementia with the transition to long-term care

A number of factors have been found to influence the transition to long-term care for people with dementia. To improve individuals' experiences of this transition, it is important that changes be made at all stages of the transition (e.g., during the decision-making, before the move, and after the move to long-term care) and that these changes be made at the individual, interpersonal, and systemic level [25]. A recent systematic review article [26] identified 19 articles that examine factors that impact residents' experience of their transition to long-term care. One factor that appears to be important is whether or not the person with dementia had a say in decisions related to their placement in long-term care [25–27]. Whether the transition was planned or an abrupt change for the person with dementia is also an important factor that can influence their experience of the transition. Individuals who are not consulted about the move will have more difficulties with the transition. In addition, the living situation of the person with dementia before the move is also a factor. For example, if they had been living alone for a while and are feeling isolated, the move will be received more positively.

Unlike people's homes, long-term care facilities are mostly depersonalized with numerous security measures and regulations in place. Although these are important for the safety of people with dementia, they also make this transition more difficult. Some research suggests providing long-term care staff with training on how to identify and provide support to individuals distressed by the transition [26]. Long-term care staff and family should monitor the person with dementia for signs of distress, anxiety, and depression after the move. In many long-term care facilities across Canada and the world (e.g., the Continuing Care Reporting System The Resident Assessment Instrument-Minimum Data Set (RAI-MDS) is used in most public long-term care facilities across Canada), individuals are assessed on a number of measures including depressive symptomatology and anxiety. This information should be monitored more closely and used to identify individuals who are having difficulties adjusting to their new home.

Living in a long-term care is often socially isolating for people with dementia in that visits from friends and family might be less frequent and participation in leisure activities more challenging. For residents, social disengagement has been associated with increased rates of depression [28], responsive behaviors [29,30], and mortality [31]. With the increased prevalence of dementia, policymakers, families, clinicians, administrators, researchers, and residents are all convinced that there are better solutions to make long-term care a place where people with dementia are safe, where relationships are developed and where residents can have something meaningful to do, despite their functional and cognitive limitations [32]. Long-term care facilities have to provide more opportunities that facilitate social connections, participation in meaningful activities, and a sense of purpose [25,26].

One solution is to include volunteers in long-term care who are specifically instructed to engage in leisure activities with the residents and who, through consistently interacting with the same resident, come to understand their personal likes and dislikes, their needs, and triggers of behaviors related to dementia [33]. This aligns with the Alzheimer Society of Canada (2011) Guidelines for Care: Person-centered care of people living with dementia in care homes [34]. Previous research on volunteer and intergenerational programs with people living with dementia have shown positive effects on well-being, quality of life, perceptions of aging, and engagement [35–37]. Our own work, as well as many others, has shown that altering social and physical environments can contribute to diminishing challenging behaviors related to dementia [30,38].

A recent intergenerational program called Recherche sur le Vieillissement et l'Intégration du Vécu en Résidence (REVIVRE) showed promising results [33]. This volunteer program matches university students with people with dementia living in long-term care. Each resident receives four 2-h visits per week by two volunteers. Therefore, as a result of this program, people with dementia living in long-term care receive support and help 4 days per week. The fact that each resident receives an additional 8 hours a week of one-on-one support over 4 days, over an extended period is what makes this program unique from other volunteer programs in long-term care. It increases the level of support and help for each resident quite extensively, thereby allowing them to develop friendships and engage in activities they may not otherwise have the chance to participate in (e.g., going outside). This program also provides caregivers an opportunity to get some much needed respite and students an invaluable learning opportunity and a changed perspective of older adults [33]. By including people with dementia who recently moved to long-term care, this program could help improve the transition (e.g., new residents would remain socially connected and communication between the family, staff, and person with dementia would be facilitated with the help of the volunteer). This program also has the potential to expand to many communities across the world.

Research also highlights the importance of autonomy for people living with dementia after the placement in long-term care home [25,26,39]. Not only do people living with dementia need to adjust to no longer living in their familiar home but they also need to get used to the loss of privacy, the depersonalization of the home, and the many security measures and regulations of the long-term care home. Some autonomy within the long-term care home can go a long way to easing this transition. Furthermore, allowing people with dementia to decorate their rooms with personal possessions is also recommended [25,26,39].

For many, the transition to long-term care is further complicated by cultural and language differences. Communication is essential to the delivery of health services, and official language minority communities are at a particular disadvantage when they access and receive services in the majority language (e.g., English-speaking residents in French speaking long-term care homes and French speaking residents in English-speaking long-term care homes). The quality of communication and language and cultural sensitivity are at the heart of each interpersonal relationship involving those who access these services and is especially important to individuals who are experiencing severe cognitive decline. Furthermore, the limited meal alternatives that ignore cultural preferences can make the transition that much more difficult. Culturally specific food preferences would help ease transitions to long-term care [26], would make the new environment feel more familiar, and might help reduce challenging behaviors related to dementia. Person-centered care that takes one's cultural background and preferences into account can improve the transition and adjustment to the long-term home [26]. The aforementioned volunteer program (i.e., REVIVRE) where French- and English-speaking university students with different cultural backgrounds volunteer in long-term care would address this issue directly, thus improving the transition for people with dementia and their caregivers.

Factors that influence the experience of the transition into long-term care by caregivers

Similar to people living with dementia, the challenges faced by caregivers start long before the move to long-term care. A recent scoping review paper by Merla et al. [19] and a systematic review by Afram et al. [11] highlighted the experiences lived by caregivers throughout the entire transition (i.e., before, during, and after the change in living arrangement).

There are a number of factors identified in the literature that can influence a caregiver's experience before the move. One important factor that can influence a caregiver's experience of the transition to long-term care is whether they have the opportunity to plan for the transition or not [19]. Some caregivers have time to contemplate and plan for the transition, whereas many others are forced into it as a result of an abrupt change in the person living with dementia's health (e.g., sudden deterioration) or the caregiver's situation. Being able to plan for the transition more gradually and making an informed decision (e.g., being able to visit long-term care homes) is associated with more positive outcomes for caregivers [19]. For some individuals, a place opens up in a long-term care home at the perfect moment whereas others are not so fortunate. For example, caregivers in the Dementia Transitions Study mentioned

> *I had been praying and praying and it was just as though God put this in my lap.*

> *Oh yea, I felt so grateful, like it just um, yea I just, I don't know why there were two openings right when we needed them, but they were there, we didn't need two but you know it was, yea I don't know. We were just luck — you know, lucky.*

Access to information from general practitioners about the dementia journey and the sometimes inevitable need to move a family member to a long-term care is also important not only because it helps to make an informed decision but also because it can help to relieve some of the guilt felt by family caregivers when deciding to place their loved ones in long-term care [19]. For example, caregivers in the Dementia Transitions Study mentioned,

> *When Dr. X said that he needed more care, that was when I met with the social worker, and she went through everything with me.*

> *We have a nurse now that visits us, and she comes and checks his blood pressure. And she came a couple of weeks ago, and she was saying 'You know <name>, now you've go to think about nursing plan for him, because, you know, things are going to get worse.*

However, the information is not always well communicated by health professionals and the stressful nature of this transition makes it that much more difficult to assimilate the information that is passed on. For example, one caregiver said

> *The problem is a lot of the language comes across as jargon, and even yourself you're asking me to kind of differentiate levels of care ... it's a confusing system when it's just being thrown at you.*

The phase before the actual move can be extremely difficult for caregivers who can be left feeling culpability, uncertainty, apprehension, anger, stress, depression, pressured, powerlessness, and worry [19]. One caregiver from the Dementia Transition Study mentioned,

> *I really don't want to move her, because every time there is a change, you know, whether it is in staff, or whether it is in people that live near her, sit at her table. It impacts her, and it just kind of puts her off a bit, so the less change I need to do the better.*

Another caregiver recounted that the person with dementia was

> *filled with fear about having to live in a care facility, and she was just filled with fear about what was going to happen to her, as her dementia progresses. And it just spiraled her down mentally.*

Even well-planned transitions can result in the need to make sudden decisions, which can be distressing to caregivers [40]. For example, caregivers are often required to put their loved ones on a waiting list when applying for a place in long-term care. What makes it difficult for caregivers is the wide variability in the length of time on the waiting list. Some individuals are on the list for more than 2 years, whereas others are assigned to a home within only a few weeks. For those expecting a wait of 2 years, having to accept a place within only a few weeks or months can be extremely distressing; however, declining the spot has major consequences. As identified by caregivers in the Dementia Transition Study, difficulties with planning the transition often arise given the complexity and unpredictability of the dementia trajectory which is further complicated by the unknown amount of time on the waiting list. For example,

> *Now that I'm seeing these changes, and they seem to be happening quite quickly here, and I'm thinking oh gosh. Ok, so I'm starting to actually look at other facilities, higher end care type facilities.*
>
> *But then I was like saying there is a 2 year wait for a lot of places. At what point do we know to get her into a senior's home as opposed to a nursing home, where she has some help? We just don't know, we are kind of overwhelmed. My dad was easy, he had Parkinson's and at one point we could say it is too much, it's going to kill mom. So we knew when it was time for him, but for Mom it's going to be a hard decision.*

The information about the admission process needs to be more carefully communicated to caregivers when they are deciding on whether or not to put their loved one on a waiting list to reduce their level of anxiety if a spot does open up quickly [40]. Often, family members are left feeling pressured and rushed by the system with no real control over when to move their family member to long-term care [19]. Support from health practitioners and social networks is important when going through this transition as those with access to more support will adjust better to the transition [19].

After people with dementia have moved to the long-term care home, many of the same emotions such as apprehension, guilt, stress, depression, and loss continue to be felt by the caregivers [19]. Caregivers and people with dementia may also feel lonely because of the transition. For example, one caregiver said *"I think we're both pretty lonely, we were together for a long time. So it has been difficult to say the least"*. Caregivers' satisfaction concerning their choice of long-term care home can have an impact on their emotions after the move. For instance, caregivers that find a good home where their loved ones are receiving quality care may feel grateful and relieved, whereas those who find a home they are not satisfied with might feel more guilt and apprehension [19]. Furthermore, continued support from family and friends, positive interactions with long-term care staff, and access to support groups where they can talk about their experiences and learn from others going through the same changes can also ease the transition [19].

Research suggests that caregivers do appreciate formal support during this difficult transition [23,41,42] but that more work is still needed to better tailor services and program to the needs of caregivers. A recent systematic review by Brooks et al. [43] about psychosocial interventions for caregivers of people with dementia after they transition to long-term care identified individualized multicomponent psychosocial interventions and group interventions as potential intervention for improving caregivers' emotional well-being and reducing feelings of guilt. However, they also mention a need for more high-quality, randomized controlled trials given the limitations of the included studies.

The importance of longitudinal studies

The needs of people with dementia and risk factors of dementia-related trajectories change over time as well as during major life transitions such as when a person moves from one setting (community) to another (long-term care). To study the change experienced by individuals living with dementia, repeated measurements of data are needed. Statistical models made possible with repeated measurements within individuals including latent growth curve models, bivariate latent growth curve model, and growth mixture models are important for understanding the trajectories of individuals with dementia before and after they transition to long-term care and the impact of different risk factors on within-person change.

Latent growth curve models allow researchers to examine trajectories (i.e., intercept and slope) of people with dementia and their caregivers (e.g., depressive symptomatology, loneliness, functioning) and examine the role of various factors (e.g., gender, social support, time to plan the transition, personalized room in the long-term care home) associated with better or worse outcomes after the transition. These factors can be time-invariant, meaning that they are only included at one time (e.g., personalization of the long-term care room when individuals move into the long-term care home), or time-varying, meaning that repeated measures of predictors (e.g., social support) are included at each occasion to take into account the time-specific fluctuations of these factors in explaining outcome variables (e.g., depressive symptomatology, functioning) above and beyond the changes predicted by the general growth trajectory [44]. The bivariate latent growth curve model is an extension of the latent growth curve model, given that it estimates the trajectory of two variables simultaneously and examines how these correlate together. This model allows for the examination of the correlation between the outcome's intercepts (e.g., correlation between level of social support and depressive symptomatology when entering long-term care), slopes (e.g., do people who decline more quickly than average on their level of social support also demonstrate a higher than average increase in depressive symptomatology?), and occasion-specific residuals (e.g., is change in social support for a specific individual at a specific occasion associated with depressive symptomology at a matched occasion?) [44]. The growth mixture model allows for the identification of classes of trajectories to take into account the heterogeneity in trajectories [45,46]. These are only a few of the statistical approaches that can be used to examine this transition.

Even with access to these statistical models, few longitudinal studies examine how trajectories change after the transition to long-term care. Most studies about this transition are cross-sectional and, therefore, focus on between-person differences rather than on within-person change. This is largely due to lack of access to longitudinal data in this area of research.

The next section discusses two research studies that use longitudinal data to examine change in behaviors related to dementia (e.g., vocal disruption, physical aggression, repetitive behaviors, and restlessness) once individuals transition to a long-term care home. Dementia-related challenging behaviors are an extreme source of stress to caregivers and long-term care staff and can have an impact on the trajectories of people with dementia and their caregivers. Therefore, a better understanding of these trajectories is important when studying dementia-related transitions.

Study one

The aim of our first study was to gain knowledge about the longitudinal relationship between dementia-related challenging behaviors and cognitive functioning in long-term care homes and about the role of demographic variables and psychosocial factors (e.g., depressive symptomatology, social engagement, cognitive functioning, activities of daily living) on these trajectories [30].

The InterRAI (http://www.interrai.org), which was modified for use in Canada, was used to examine the trajectories of people with dementia once they enter long-term care homes in Ontario, Canada [47,48]. This assessment instrument includes demographic information as well as clinical and functional characteristics of the residents such as cognitive function, psychosocial well-being, health conditions, communication/hearing, behaviors, and physical function. These assessments were completed by nurses every 3 months for the duration of the resident's stay in long-term care, making these data relevant to researchers interested in understanding the trajectories of people with dementia living in long-term care homes. More detailed information about the methodology is available in the published article [30]. The InterRAI has been adopted worldwide, making it an interesting source of data for researchers around the world. The Minimum Data Set (MDS) Aggressive Behavior Scale (ABS) was used to assess frequency and intensity of residents' challenging behaviors in the last 7 days [49]. The Cognitive Performance Scale (CPS) was used to assess older adults' level of cognition [50]. The MDS Depression Rating Scale (DRS) [51] was used to assess depression symptoms of older adults in long-term care. The Index of Social Engagement (ISE) [52] was used to assess level of social engagement. The MDS (section G: Physical Functioning and Structural Problems) Activities of Daily Living—Long Form [53] was used to assess activities of daily living. The analyses were completed within the structural equation modeling framework using a multivariate latent growth model to examine both trajectories (cognitive functioning and dementia-related challenging) simultaneously. The final sample included 16,810 older adults ranging in age from 65 to 109 years.

Findings from this study suggest that challenging behaviors increase upon entry in long-term care but level off at later assessment times, highlighting the difficulties faced by people with dementia when transitioning to long-term care. Resources and person-centered approaches that take into account the person's likes/dislikes, routines, and needs would ideally be initiated before transitioning to the long-term care to help ease the transitions and decrease challenging behaviors [30]. The study findings also suggest that trajectories of cognitive functioning and behaviors are related. That is, on average, individuals who were more cognitively impaired upon entry into long-term care also exhibited more behaviors, those who exhibited a steeper increase in cognitive impairment also exhibited a steeper increase in challenging behaviors and, at the within-person level, individuals demonstrating an increase in cognitive impairment at a specific occasion were also more likely to demonstrate an increase in challenging behaviors at that same occasion. Furthermore, younger males with higher depressive symptomatology, more impairment with activities of daily living, and less social engagement tended to exhibit more challenging behaviors. These findings highlight the importance in understanding factors related to dementia-related trajectories after a move to long-term care as these can help to identify avenues for improvement and change. For example, preventing, detecting, and treating depression among people with dementia and making sure they are given opportunities to participate in social activities could help to improve the trajectories of people with dementia after the move to long-term care.

Study two

In the aforementioned study, the average trajectory of dementia-related challenging behaviors in long-term care was examined. However, this approach assumes that change in dementia-related challenging behaviors in long-term care can be described by a single trajectory. It is more reasonable to assume that there might be heterogeneity in the trajectories and that certain profiles of people with dementia cluster together. For example, it is likely that some individuals enter long-term care exhibiting few behaviors but show a fast increase in behaviors, whereas others might enter long-term care exhibiting few behaviors with no further increase over time. Recent developments in the analysis of longitudinal data have resulted in an ability to better understand heterogeneity in trajectories. One such development, referred to as Growth Mixture Modeling, allows for the identification of different classes of individuals whose trajectories show clusters of similar patterns rather than describing a homogenous trajectory within a given population [45]. By using this approach, the importance of key factors such as social engagement, depressive symptomatology, and activities of daily living can be examined more closely in terms of whether they have an impact on class specific trajectories of challenging behaviors but also on whether these factors help to discriminate individuals' membership to the different classes. For example, individuals who are more socially engaged when entering long-term care may be more likely to be in the class with a low frequency of challenging behaviors. The purpose of this study was to identify subgroups of individuals with similar developmental trajectories of challenging behaviors related to dementia in long-term care and evaluate the role of demographic variables, depressive symptomatology, social engagement, cognitive functioning, and activities of daily living on class membership.

The same data, sample, and measures were used as in the aforementioned study. Growth mixture models were run to identify unobserved groups of individuals with similar developmental trajectories of challenging behaviors related to dementia. Based on our previous work [30] where we reported that changes in behaviors follow a nonlinear trajectory, we fitted a growth mixture model that estimated curvilinear trajectories for each class. Class membership was estimated using a regression adjusted for social engagement, cognitive functioning, depressive symptomatology, and activities of daily living.

Our results demonstrate the existence of heterogeneity in trajectories of challenging behaviors for people living with dementia in long-term care that is best represented by seven classes of trajectories rather than one homogenous group. This second study extends our previous work on the trajectory of dementia-related challenging behaviors and is the first study to identify classes of trajectories of these behaviors in long-term care.

Our results demonstrate that the majority (69%) of people living with dementia entering long-term care exhibit few challenging behaviors and that these behaviors increase gradually over time. This class resembles the results found from our previous study [30]; however, this study further unravels the distinct trajectories that older adults demonstrate as they transition to long-term care, highlighting how some individuals do not adapt as well as others. For example, we found that in two of the seven classes, the individuals enter long-term care with few challenging behaviors but exhibit a steep increase in behaviors right after entering long-term care followed by a steep decrease in behaviors in later months. These individuals would benefit most from interventions aimed at easing their transition in long-term care and from more person-centered care approaches by getting to know the individual before they enter long-term care. Our results also identified a class of older adults who exhibit few behaviors upon entry in long-term care but who demonstrate a linear increase over time; one that is more pronounced than those reported in the first class. These individuals will continue to demonstrate an increase in behaviors over time unless efforts are made to address the cause of these behaviors. There is also a class of people living with dementia who exhibit the highest frequency of challenging behaviors upon entry in long-term care and who continue to exhibit a consistently high frequency of behaviors over time.

We also found that depressive symptomatology, social engagement, activities of daily living, and cognitive functioning are important in predicting class membership. This is valuable information that can be used to identify groups of individuals at increased risk of having more challenging behaviors related to dementia. It also means that initiatives put forth in the community, while people living with dementia are still living at home, could help to reduce the number of people living with dementia in the classes exhibiting more behaviors. Once individuals are in long-term care, actions can also be taken to modify an individual's trajectory of challenging behaviors.

Among all the included covariates, depressive symptomatology stands out as the most consistently associated with the trajectories of challenging behaviors. Social engagement also appears to play an important role in most trajectories of challenging behaviors. The longitudinal importance of social engagement further reinforces the importance of keeping people living with dementia who move to and live in long-term care homes engaged in activities with regular human contact [5]. The importance of tackling loneliness in older adults in long-term care is something that can never be underestimated.

The need for further longitudinal research

Further longitudinal research is still needed to help identify people with dementia at increased risk of experiencing difficulties when transitioning to long-term care and highlight interventions that could help ease these transitions. That is, the aforementioned studies have focused specifically on better understanding the trajectories of challenging behaviors related to dementia. However, many other trajectories such as social isolation, physical health, quality of life, medication use, and depressive symptomology could provide some information about the trajectories of people with dementia when they move to a long-term care home.

Furthermore, the two aforementioned studies focus specifically on the trajectories after the move to a long-term care home. However, as mentioned, the transition from living at home to living in a long-term care home is much more complex and includes a period of time before and after the change in living arrangement. More studies are needed that examine the trajectories in functioning before and after moving from one setting (home) to another (long-term care home) to see how trajectories differ across settings and to examine potential reasons for transiting across settings and improve predicted likelihood of change. Care and support for people living with dementia and their caregivers should be integrated across both levels of care provision (while living in their home and after moving to long-term care) to smooth the transition across settings. A better understanding of trajectories before and after moving to long-term care would allow care systems and families to be better prepared and to provide more person-centered support. It would also allow for the development of interventions that are better tailored to the needs of the person.

More longitudinal studies are also needed that examine the trajectories of caregivers before and after the transition to long-term care to further understand caregiver's experience of the transition and the factors that make this transition easier or harder for them specifically. Although some longitudinal data about caregivers exist, more large-scale studies are needed that follow caregivers of people with dementia through their caregiving journey over an extended period. Existing cohort studies that ask participants whether or not they are caregivers and that include information about the type and intensity of caregiving and whether or not the person they are caring for is living with them (i.e., so that researchers can identify when the caregiver goes from caring for them in the home to caring for them in a long-term care setting) could also be used. For example, in Canada, the Canadian Longitudinal study on Aging (CLSA), and, in Europe, the Survey of Health, Aging and Retirement in Europe (SHARE) include questions about caregiving across time.

Conclusion

In this chapter, I discuss one of the last transitions that people with dementia and their caregivers must go through, the transition to a long-term care home. This transition, which can last for months or even years, has an impact on the person with dementia and their family. Although a lot of progress has been made in understanding factors related to this transition, more longitudinal research is needed to better understand the trajectories of people with dementia and their caregivers when transitioning to long-term care so that people can be better counseled during these transitions and expectations can be better managed. Throughout this chapter, I refer to the long-term care living arrangement as a home (e.g., long-term care home) rather than as a facility (e.g., long-term care facility) because I believe we need to change the culture around long-term care and the perceived stigma with this living arrangement (i.e., long-term care homes need to become more like a home away from home). For most, this is their last living arrangement before their death. Why not make sure it is tailored to people's needs, likes, and dislikes and that it includes opportunities for social interaction and participation in meaningful activities.

References

[1] Ferri CP, Prince M, Brayne C, Brodaty H, Fratiglioni L, Ganguli M, et al. Global prevalence of dementia: a Delphi consensus study. Lancet 2005;366(9503):2112–7.
[2] World Health Organization and Alzheimer's Disease International. Dementia: a public health priority. 2012. Geneva, Switzerland.
[3] Prince M, Bryce R, Albanese E, Wimo A, Ribeiro W, Ferri CP. The global prevalence of dementia: a systematic review and metaanalysis. Alzheimer's Dement 2013;9(1):63–75.e2.
[4] Alzheimer's Association. Alzheimer's disease facts and figures. Alzheimer's Dement 2016;12(4):459–509. 2016.
[5] Livingston G, Sommerlad A, Orgeta V, Costafreda SG, Huntley J, Ames D, et al. Dementia prevention, intervention, and care. Lancet 2017;390(10113):2673–734.
[6] Rose KM, Lopez RP. Transitions in dementia care: theoretical support for nursing roles. Online J Issues in Nurs 2012;17(2):4.
[7] Sanford AM, Orrell M, Tolson D, Abbatecola AM, Arai H, Bauer JM, et al. An international definition for "nursing home". J Am Med Dir Assoc 2015;16(3):181–4.
[8] Walker CA, Curry LC, Hogstel MO. Relocation stress syndrome in older adults transitioning from home to a long-term care facility. J Psychosoc Nurs Ment Health Serv 2007;45(1):39–45.
[9] Manion PS, Rantz MJ. Relocation stress syndrome: a comprehensive plan for long-term care admissions: the relocation stress syndrome diagnosis helps nurses identify patients at risk. Geriatr Nurs 1995;16(3):108–12.
[10] Sinha S. Living longer, living well: report submitted to the minister of health and long-term care and the minister responsible for seniors on recommendations to inform a seniors strategy for Ontario. 2012.
[11] Afram B, Verbeek H, Bleijlevens MH, Hamers JP. Needs of informal caregivers during transition from home towards institutional care in dementia: a systematic review of qualitative studies. Int Psychogeriatr 2015;27(6):891–902.
[12] Collins C, Stommel M, Wang S, Given CW. Caregiving transitions: changes in depression among family caregivers of relatives with dementia. Nurs Res 1994;43(4):220–5.
[13] Etters L, Goodall D, Harrison BE. Caregiver burden among dementia patient caregivers: a review of the literature. J Am Acad Nurse Pract 2008;20(8):423–8.
[14] Lévesque L, Ducharme F, Lachance L. A one-year follow-up study of family caregivers of institutionalized elders with dementia. Am J Alzheimers Dis 2000;15(4):229–38.
[15] Schulz R, Belle SH, Czaja SJ, McGinnis KA, Stevens A, Zhang S. Long-term care placement of dementia patients and caregiver health and well-being. Jama 2004;292(8):961–7.
[16] Aminzadeh F, Dalziel WB, Molnar FJ, Garcia LJ. Meanings, functions, and experiences of living at home for individuals with dementia at the critical point of relocation. J Gerontol Nurs 2010;36(6):28–35. quiz 6-7.
[17] Aminzadeh F, Dalziel WB, Molnar FJ, Garcia LJ. Symbolic meaning of relocation to a residential care facility for persons with dementia. Aging Ment Health 2009;13(3):487–96.
[18] Drummond N, McCleary L, Garcia L, McGilton K, Molnar F, Dalziel W, et al. Assessing determinants of perceived quality in transitions for people with dementia: a prospective observational study. Canad Geriat J 2019;22(1):10. 2019.
[19] Merla C, Wickson-Griffiths A, Kaasalainen S, Bello-Haas VD, Banfield L, Hadjistavropoulos T, et al. Perspective of family members of transitions to alternative levels of care in anglo-saxon countries. Curr Gerontol Geriatr Res 2018;2018.

[20] Garner R, Tanuseputro P, Manuel DG, Sanmartin C. Transitions to long-term and residential care among older Canadians. Health Reports 2018;29(5):13–23.
[21] Caron CD, Ducharme F, Griffith J. Deciding on institutionalization for a relative with dementia: the most difficult decision for caregivers. Canad J Aging 2006;25(2):193–205.
[22] Ducharme F, Couture M, Lamontagne J. Decision-making process of family caregivers regarding placement of a cognitively impaired elderly relative. Home Health Care Serv Q 2012:31.
[23] Butcher HK, Holkup PA, Park M, Maas M. Thematic analysis of the experience of making a decision to place a family member with Alzheimer's disease in a special care unit. Res Nurs Health 2001;24(6):470–80.
[24] Lord K, Livingston G, Cooper C. A systematic review of barriers and facilitators to and interventions for proxy decision-making by family carers of people with dementia. Int Psychogeriatr 2015;27.
[25] Sussman T, Dupuis S. Supporting residents moving into long-term care: multiple layers shape residents' experiences. J Gerontol Soc Work 2014;57(5):438–59.
[26] Brownie S, Horstmanshof L, Garbutt R. Factors that impact residents' transition and psychological adjustment to long-term aged care: a systematic literature review. Int J Nurs Stud 2014;51(12):1654–66.
[27] Lord K, Livingston G, Robertson S, Cooper C. How people with dementia and their families decide about moving to a care home and support their needs: development of a decision aid, a qualitative study. BMC Geriatr 2016;16(1):68.
[28] van Beek AP, Frijters DH, Wagner C, Groenewegen PP, Ribbe MW. Social engagement and depressive symptoms of elderly residents with dementia: a cross-sectional study of 37 long-term care units. Int Psychogeriatr/IPA 2011;23(4):625–33.
[29] Cohen-Mansfield J, Marx MS, Werner P. Agitation in elderly persons: an integrative report of findings in a nursing home. Int Psychogeriatr 1992;4(Suppl. 2):221.
[30] Robitaille A, Garcia L, McIntosh C. Joint trajectories of cognitive functioning and challenging behavior for persons living with dementia in long-term care. Psychol Aging 2015;30(3):712–26.
[31] Kiely DK, Simon SE, Jones RN, Morris JN. The protective effect of social engagement on mortality in long-term care. J Am Geriatr Soc 2000;48(11):1367–72.
[32] Baines D, Armstrong P. Promising practices in long term care: ideas worth sharing. Ottawa: Canadian Centre for Policy Alternatives; 2016.
[33] Garcia L, Robitaille A, Lacelle C, Egan M, Savard J, Lemay G, et al. REVIVRE! Research on aging and integration of life experiences in residence: encouraging university students to work with seniors in long-term care homes. Alzheimer's Dement 2016;12(7):P1127.
[34] Alzheimer Society of Canada. Guidelines for Care: person-centred care of PWD living in care homes Toronto (ON). Alzheimer Society of Canada; 2011.
[35] Galbraith B, Larkin H, Moorhouse A, Oomen T. Intergenerational programs for persons with dementia: a scoping review. J Gerontol Soc Work 2015;58(4):357–78.
[36] Jarrott SE, Bruno K. Intergenerational activities involving persons with dementia: an observational assessment. Am J Alzheimer's Dis Other Dementias 2003;18(1):31–7.
[37] Blais S, McCleary L, Garcia L, Robitaille A. Examining the benefits of intergenerational volunteering in long-term care: a review of the literature. J Intergener Relat 2017;15(3):258–72.
[38] Cohen-Mansfield J, Thein K, Marx MS, Dakheel-Ali M, Murad H, Freedman LS. The relationships of environment and personal characteristics to agitated behaviors in nursing home residents with dementia. J Clin Psychiatry 2012;73(3):392.

[39] Johnson RA, Bibbo J. Relocation decisions and constructing the meaning of home: a phenomenological study of the transition into a nursing home. J Aging Stud 2014;30:56–63.
[40] Caldwell L, Low LF, Brodaty H. Caregivers' experience of the decision-making process for placing a person with dementia into a nursing home: comparing caregivers from Chinese ethnic minority with those from English-speaking backgrounds. Int Psychogeriatr 2014;26(3):413–24.
[41] Bramble M, Moyle W, McAllister M. Seeking connection: family care experiences following long-term dementia care placement. J Clin Nurs 2009;18(22):3118–25.
[42] Kelsey SG, Laditka SB, Laditka JN. Caregiver perspectives on transitions to assisted living and memory care. Am J Alzheimer's Dis Other Dementias 2010;25(3):255–64.
[43] Brooks D, Fielding E, Beattie E, Edwards H, Hines S. Effectiveness of psychosocial interventions on the psychological health and emotional well-being of family carers of people with dementia following residential care placement: a systematic review. JBI Database System Rev Implement Rep 2018;16(5):1240–68.
[44] Muniz-Terrera G, Robitaille A, Kelly A, Johansson B, Hofer S, Piccinin A. Latent growth models matched to research questions to answer questions about dynamics of change in multiple processes. J Clin Epidemiol 2017;82:158–66.
[45] Muthén B, Shedden K. Finite mixture modeling with mixture outcomes using the EM algorithm. Biometrics 1999;55(2):463–9.
[46] Ram N, Grimm KJ. Growth mixture modeling: a method for identifying differences in longitudinal change among unobserved groups. Int J Behav Dev 2009;33(6):565–76.
[47] Hirdes JP, Ljunggren G, Morris JN, Frijters DHM, Finne Soveri H, Gray L, et al. Reliability of the interRAI suite of assessment instruments: a 12-country study of an integrated health information system. BMC Health Services Research 2008;8:277.
[48] Hirdes JP, Fries BE, Morris JN, Steel K, Mor V, Frijters D, et al. Integrated health information systems based on the RAI/MDS series of instruments. Healthc Manag Forum 1999;12(4):30–40.
[49] Perlman CM, Hirdes JP. The aggressive behavior scale: a new scale to measure aggression based on the minimum data set. J Am Geriatr Soc 2008;56(12):2298–303.
[50] Morris JN, Fries BE, Mehr DR, Hawes C, Phillips C, Mor V, et al. MDS cognitive performance Scale©. J Gerontol 1994;49(4):M174–82.
[51] Burrows AB, Morris JN, Simon SE, Hirdes JP, Phillips C. Development of a minimum data set-based depression rating scale for use in nursing homes. Age Ageing 2000;29(2):165–72.
[52] Mor V, Branco K, Fleishman J, Hawes C, Phillips C, Morris J, et al. The structure of social engagement among nursing home residents. J Gerontol 1995;50B(1):P1–8.
[53] Morris JN, Morris SA, Fries BE. Scaling ADLs within the MDS. J Gerontol 1999;54(11):M546–53.

CHAPTER 9

A palliative approach to care: from diagnosis to end-of-life

Genevieve Thompson[1], Abigail Wickson-Griffiths[2]
[1]Associate Professor, Nursing, University of Manitoba, Winnipeg, MB, Canada; [2]Assistant Professor, Nursing, University of Regina, Regina, SK, Canada

Improving the quality of life and the quality of palliative and end-of-life care for people living with dementia has garnered increasing attention in both the research literature and in clinical practice. Using the principles of a palliative approach to care for those dying with, but not necessarily from, dementia has been identified as valuable and important [1,2]. However, historically, few people with dementia receive this type of care [3]. Barriers to adopting a palliative approach have been cited in the literature [4] and may stem from three factors. First, dementia is typically not acknowledged as a terminal illness. Second, palliative care is often equated with end-of-life care and care of those actively dying. Third, people with dementia lose their ability to verbally communicate, resulting in challenges in engaging in goals of care discussions, advance care planning, and pain and symptom identification. Despite these challenges, the successful adoption of a palliative approach to care is possible for people with dementia and their care partners if and when the broader understanding of what constitutes a palliative approach is adopted. Drawing on the recommendations for a palliative approach to dementia care which views palliative care as more than just care of a person who is at the end of life and is actively dying [2], this chapter will present the lived experience of people with dementia and their care partners to illustrate key ingredients of a successful transition at the final stage of life.

Acknowledging the terminal nature of dementia

To effectively provide a palliative approach for people living with dementia and their care partners, it must be acknowledged that dementia is a progressive, terminal disease. In several studies, when questioned about whether dementia was a disease you could die from, few participants agreed [5,6]. The experience of family care partners reported in the research

Evidence-informed Approaches for Managing Dementia Transitions
ISBN 978-0-12-817566-8
https://doi.org/10.1016/B978-0-12-817566-8.00009-7

literature reflects this perspective; few had an understanding of the clinical course of dementia or that it was life-limiting [7–10]. As stated by a care partner in the study by Andrews and colleagues [10]: "... *I don't think she [Mum] will die from dementia, I think she will die from a heart attack or stroke ... some other medical condition but not dementia ... do people die from dementia? I've never ... heard of people dying [from it]* (pg. 566)".

Typically, dementia is characterized as a progressive disease with an unpredictable course of illness, a "steady prolonged dwindling" with no discernible end stage [11]. Similar to the frailty trajectory [12], dementia progression is characterized by a much lower level of functioning during the last year of life, with more severe decline during the last month before death [13]. In general, the trajectory of dementia is often delineated into mild, moderate, and severe corresponding to early, mid, and late (or advanced) stages of disease [14]. While staging is helpful in underscoring the corresponding changes in functional abilities of people with dementia, determining the prognosis for people living with dementia has been described as highly inaccurate and time to death is difficult to predict [15]. While having an accurate sense of when an individual may be approaching the end of their life may assist in identifying end-stage dementia and therefore may increase access to formal palliative care options that are often predicated on the eligibility for benefits (e.g., 6 months or less to live to qualify for hospice) [16,17], it may also be problematic. If palliative care is equated with end-of-life care and is only implemented in the advanced stages of the illness, a stage when most people with dementia have lost their ability to verbally communicate; people with dementia may be excluded from participating in decisions about the care provided to them. Many individuals will die before reaching the advanced stages of dementia [18] and therefore will not reap the benefits of a palliative approach to the disease.

Healthcare providers similarly are challenged in recognizing dementia as a terminal illness [19,20]. Robinson et al. [21] identified that although dying with dementia is common in residential nursing homes, approximately 50% of healthcare providers lacked knowledge regarding the life-limiting nature of dementia and the symptoms associated with disease progression. In studies examining death certificates, dementia is rarely listed as a cause of death [22]. There are a number of reasons why dementia may not be readily viewed as the cause of, or even contributing to, a person's death. Cox and Cook's [23] research illustrating the three ways of dying with dementia is helpful in this regard: (1) those who reach the end of their lives with dementia but die as a result of another identified condition (e.g., cancer, heart disease); (2) those who reach the end of life with a mixture of physical and mental conditions, but cognitive impairment is not advanced; and (3) those who die as a result of the complications of dementia.

There also appears to be a misunderstanding among care partners of the contribution of progressive cognitive decline toward other commonly experienced medical complications such as immobility, challenges in swallowing, recurrent infections, and ultimately death [10]. In general, family care partners tend to focus on the immediate day-to-day changes they see in people with dementia, rather than the bigger picture and the fact that these changes such as swallowing difficulties may be indicative of decline toward death [11,24]. Study findings note that family care partners of people with dementia were shocked when death occurred as they had not recognized their relative was dying: "*... I was also totally in shock because I wasn't expecting her to die so fast and I don't think I was prepared for that personally*" [25] (pg. 334) *and "So, when she did finally die it was kind of a bit of a shock to us really ... we hadn't a clue that that was going to happen"* [26] (pg. 14).

The failure to recognize dementia as a terminal illness can result in adopting practices that may not achieve high-quality palliative care. Past research has underscored that people with dementia are at high risk of poor outcomes at the end of life such has being significantly less likely to have advance directives limiting aggressive care (such as tube feeding), emergency department visits, hospitalizations, and inadequate assessment of pain and dyspnea. They experience higher rates of pressure ulcers, pneumonia, and fever than patients with cancer [3,13,27]. When family care partners understand the clinical trajectory of dementia, care is improved [5,28], specifically in higher levels of comfort for people with dementia owing to a decrease in use of burdensome interventions, and fewer diagnostic tests and therapeutic treatments. By accepting that dementia is a life-limiting illness and adopting a palliative approach to care from diagnosis of the disease, people with dementia and their care partners will benefit from having their needs placed at the center of care. Educating care partners about the clinical trajectory could result in better knowledge about disease progression, thus influencing care decisions. This is illustrated by the following description provided by a family care partner:

> *I don't think some of the stuff you need to know is easily available and I think the more anybody can do to further the knowledge and the understanding so you*

get there quicker and sort things out as soon as you can to give that person the best quality of life they can have [the better] [29] (pg. 1599).

Adopting a palliative approach to care for people with dementia

As evident in preceding chapters, people with dementia have complex and multifaceted needs that require different healthcare and social care services at junctures along their disease trajectory. Adopting an approach tailored to meet these needs is critical to the delivery of high-quality care. A palliative approach to care is particularly relevant to people with dementia [2,30]. This adapts and integrates the core principles and values of palliative care, such as a focus on the quality of life of the person, patient- and family-centered care, and impeccable pain and symptom management, into the care received by persons who have life-limiting conditions but are not yet at their final stages [31]. The palliative approach to care can be implemented alongside chronic disease management [32]. A palliative approach does not dichotomize care or force individuals to choose between "curative" vs. "comfort" care. Rather, palliative care is offered from the time of diagnosis and throughout the illness, along with acute or curative management as appropriate according to need. As those needs change over time and a curative intent declines, comfort care takes a more prominent role [32]. As such, end-of-life care is part of a palliative approach to care but is not the sole focus. Fig. 9.1 illustrates elements of palliative care, end-of-life care, and care in the last hours of life.

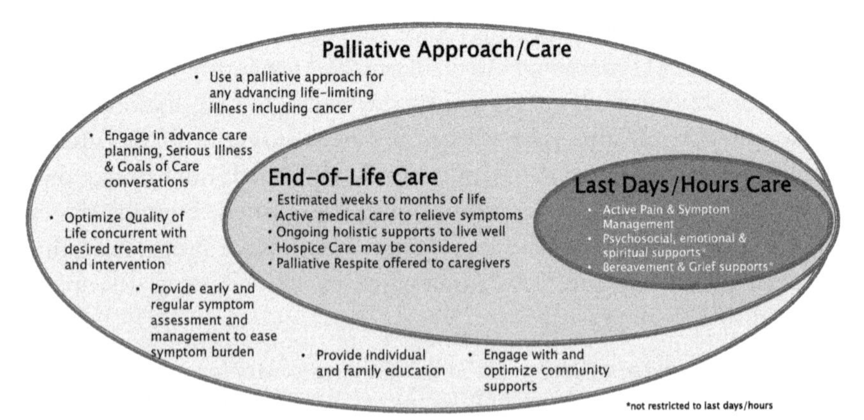

Figure 9.1 Differentiating and understanding the palliative continuum: palliative approach/care vs. end-of-life care vs. last days/hours care *(Source: BC Palliative Centre for Excellence, June 26th, 2013. Updated: Interior Health, July 2019).*

The palliative approach was first described by Kristjanson et al. [33] as an approach to care for those with advancing chronic illness who may not require specialized palliative care services such as hospice care yet would benefit from healthcare providers who espouse an open and positive attitude toward death and dying. In this approach, healthcare providers are equipped with the tools for the prevention and relief of suffering through early identification, assessment, and treatment of physical, psychosocial, spiritual, and existential problems. Sawatzky et al. [34] have further identified three key themes relevant to a palliative approach to care, including: (1) an upstream orientation to ensure that the needs of individuals with a life-limiting condition and their family are addressed early (e.g., from time of diagnosis); (2) the adaptation of palliative care knowledge and expertise to the specific needs of the population in question; and (3) the integration and conceptualization of the palliative approach into every relevant setting of care.

In using this approach with people living with dementia, the European Association of Palliative Care [2] postulated that to acknowledge the importance of a palliative approach across the dementia spectrum, one must recognize that care goals will change depending on the needs of individuals and their care partners. The "dose" of palliative care may fluctuate in its intensity along the dementia course, including the level of support required by the care partner [35]. For example, a husband and daughter took care of their relative living with dementia for 5 years at home without assistance, but approximately 3 months before her mother's death, the daughter said to her father, "*this is not working, and you need more help*" [36] (pg. 32). After this, the family enrolled the mother with hospice care where she received a necessary mobility aide as well as personal and nursing care. This approach resonates with many care partners who have identified that they experience the end-of-life period for significantly longer than the final 12 months of life, and many express a readiness for a palliative approach from several years earlier [37,38]. Adopting a palliative approach from the time of diagnosis may, therefore, alleviate unnecessary suffering for both the individual with dementia and their care partners.

Importance of decision-making and advance care planning

Early acknowledgment of the terminal nature of dementia also allows for decisions regarding treatments and supports to occur within the larger context of the disease rather than at a time of crisis [39,40]. In recognizing that people with dementia will lose their capacity to make informed care decisions in the later stage of the disease, advance care planning is essential in the continuing management of dementia [13]. As the disease progresses, responsibility for decision-making shifts from the people with dementia themselves to their family and professional care partners, highlighting the importance of advance care planning. By definition:

> *Advance care planning is a process that supports adults at any age or stage of health in understanding and sharing their personal values, life goals, and preferences regarding future medical care. The goal of advance care planning is to help ensure that people receive medical care that is consistent with their values, goals and preferences during serious and chronic illness* [41] (pg. 821).

From the perspective of people with dementia, advance care planning is the strongest and the most consistent modifiable factor in helping to avoid treatments or transitions that are not desired or necessary [42]. While the evidence base is not robust, advance care planning is linked to positive end-of-life outcomes, including preferred place of death, treatment, satisfaction with care, and lower physical or emotional distress [43].

Even though the value of advance care planning for people with dementia is commonly recognized, it does not happen routinely [44]. Indeed, few will have the chance to engage in advance care planning, as the time of the diagnosis may seem too early for such discussions, the person with dementia may worry about their healthcare wishes limiting the future choices of their care partners, their cognitive capacity is viewed by care partners and healthcare professionals as a barrier, or advance care planning is perceived to be an upsetting experience [43–45]. However, these assumptions may be false. Care partners of people with dementia living in long-term care homes often regret not having had advance care discussions [46]. Such discussions are feasible. With support, people with dementia and their care partners can engage in and have positive experiences with advance care planning. Much of the research about advance care discussions has been conducted in long-term care homes, when the person with dementia is at a later stage of illness, so we do not have good evidence about the best ways to support advance care discussions. However, conversation guides are available and are being tested in ongoing research. For example, *Your Conversation Starter Kit for Loved Ones of People with Alzheimer's Disease or Other Forms of Dementia* is available online through The Conversation Project in English, French and Spanish.

Another barrier to advance care planning for people with dementia is knowledge and confidence of healthcare professionals who may not recognize the benefit of advance care planning for this population and may lack skills needed to initiate and document discussions and implement care [47]. Piers et al. [44] found that high-quality, evidence-based advance care planning guidelines targeted for people with dementia are not available, which also may contribute to lower uptake of advance care planning. Together, these barriers may limit the opportunities to incorporate a palliative approach for care. To support healthcare providers, Piers et al. [44] developed 32 recommendations for advance care planning for people with dementia and their family care partners, based on literature as well as contributions from experts, end users, and peers. Although based on low to very-low levels of evidence, the recommendations for practice provide guidance. The recommendations fall into eight domains including (1) initiating advance care planning; (2) evaluating mental capacity; (3) performing advance care planning conversations; (4) the role and importance of persons close to the people with dementia; (5) advance care planning when verbal communication is no longer possible; (6) documenting wishes and preferences; (7) end-of-life decision-making; and (8) preconditions for optimal implementation. According to these guidelines, not only should advance care planning be initiated by healthcare providers soon after diagnosis but also it should be scheduled as part of routine practice and also occur spontaneously as opportunities for discussion present. Healthcare providers should watch for triggers for ongoing planning such as change in health status. While healthcare providers should assess mental status and capacity, the guidelines caution that one should assume that the person is capable of discussing their preferences and wishes and remember that mental capacity fluctuates. The guidelines include 10 recommendations about how to have advance care discussions, including some about communication skills such as adapting conversation to the person's abilities, being relaxed and natural in the conversation, listening to understand the person's values, preferences and expectations in context of their life story, listening to the persons' fears and concerns, not forcing discussion with a person who is reluctant, and breaking conversations up by discussing different topics on different occasions. The content of advance care planning should include discussion of who else should be included in the discussion; clarifying misunderstanding about dementia trajectory;

discussing values, beliefs, preferences, and expectations in relation to the present and future; and specifying care goals and end-of-life preferences.

Treatment preferences

Advance care planning discussions ultimately support goal-setting and decision-making, in anticipation of the person with dementia no longer being able to make informed treatment decisions. The process may also be initiated within a "best interest" framework with care partners, when the individual is no longer able to communicate [43]. Successful discussions about treatment preferences require that the person with dementia and care partners understand the clinical course of dementia.

A central focus of goal-setting is medical treatment for common complications and distressing symptoms that persons with advanced dementia may experience. Common complications include the reduced ability to swallow or fight infections that contribute to or cause death [42]. Treatable distressing symptoms that may be addressed include shortness of breath, pain, and pressure ulcers [28]. However, advance care planning should be holistic in nature, extending beyond medical treatments by considering overarching life goals or priorities [41]. Goals of care for medical treatment and support can be considered along on a continuum—from receiving treatment to limit discomfort through to life-sustaining interventions [42]. Within this continuum, people with dementia whose goals are consistent with care focused on comfort may also share preferences about receiving curative but noninvasive or minimally invasive treatment to alleviate acute episodic ailments (e.g., oral antibiotic to treat infection) to maintain their baseline health status. Within the context of a palliative approach, goals for comfort and curative intervention are not mutually exclusive [42]. However, as recognized by van der Steen et al. [2], as death becomes more imminently anticipated, the individual's comfort should be maximized.

Considering implications for treatment preferences—place of care

In considering a holistic approach to advance care planning, discussions should include a person's preference for place of care [48], which may ultimately be influenced by their treatment preferences (e.g., some treatments or care options may not be available in the community, at home or in a nursing care facility). Certainly, discussions about the place of care may also help to limit burdensome admissions to acute care settings [43]. Importantly, research indicates that if adequately supported, most people wish to die in their own home, as it may bring about feelings of comfort and security. Yet people with dementia typically die in a hospital or long-term care setting [36,48]. During advance care planning, when there is a preference to remain home, little attention may be given to the implications for care partners and availability of support services [36]. However, considering family and care partner perception is an important element in the palliative approach to care and their safety and potential for burden should be explored in the decision-making [48]. As discussed in chapters 7 and 8, as dementia progresses, it may become impossible to safely stay at home, necessitating a move to an assisted living or long-term care home often well before the person's death. People come to view these settings as home and prefer them as a place of death over hospital. A palliative approach in long-term care homes may reduce hospital admissions before death [49].

Family care and involvement

As indicated by van der Steen et al. [2] and the palliative approach, family care partners' involvement may be encouraged throughout a person with dementia's journey, regardless of care setting. As end-of-life approaches, family care partners may have a strong caring relationship with the person with dementia, but they may struggle with burden, fatigue, anticipatory grief, as well as feelings of isolation and guilt [26,37]. In addition, at the end of life, family care partners can experience feelings of grief and need for bereavement care, marking yet another critical transition as they relinquish caregiving responsibilities [2,25]. Overall, literature indicates that through the caregiving experience, care partners develop physical, emotional, and psychological needs, as well as needs related to decision-making and instrumental support [35]. This remains true through the late stages of dementia and the person's death. Care partners can require support healthcare providers as well as others in their family and social networks to navigate challenges and transitions from the time of diagnosis through to the final stages of illness and into the bereavement period [9].

Even in advanced stages of dementia, family care partners may not fully comprehend the dementia trajectory and need explanations about the implications of decline and, ultimately, death. Research with care partners of people with advanced dementia living in long-term care homes consistently finds that they have questions about what to expect as the person deteriorates and about what death will be like [46,50]. Furthermore, they find this information reassuring [51]. However, care partners often find that healthcare providers in long-term care homes and other settings are reluctant to discuss death with them [46,50,52]. This is illustrated in this statement by a care partner in a long-term care home:

> *To me, helping people understand, really, really understand what an end of life process is. It just. Then there still seems to be an incredible amount of reluctance on the provider's side to even talk about that. Because it's not necessarily a pretty process. It seems like the general belief is it's better for people to be surprised by it or whatever. I don't know what the hell it is but and because of that people don't properly prepare* [46] (pg. 6).

Caring in the community

For people with dementia who express the desire to remain "at home" during their end-of-life period, this means a significant commitment from their family care partners, who may require community-based healthcare services and informal supports to manage the personal and practical caregiving challenges [25,48]. It is more likely that people with dementia will remain at home to receive end-of-life care should their care partner have personal attributes such as resilience and the determination to provide such care; the ability to recognize and manage symptoms; and access to informal supports from others to help them cope with their role (e.g., family, friends, volunteers) [48]. While the care partner relationship can be positively perceived as bringing those with dementia and their partners closer together, the findings of Dempsey et al. [26] highlight the practical and psychosocial support needs related to care partner involvement. In that study, a daughter of a person with late stage dementia explained:

> *the focus has to be on the individual, it has to be on the person. [ ] All our lives were just totally, focused around mum and dad prior to that. And you had no time for anybody else and any time you had for somebody else, you were feeling so guilty that you should have being given it that way. You're so divided. (pg. 10).*

The son of another person with dementia explained:

> *And even with the medication now they say to me well it's at your own discretion, you can up it if you like and I said no, I couldn't do that, I can't. I have enough responsibility [ ]. I'm not going to start prescribing medication here as well, I can't do that.* (pg. 7).

These examples illustrate how care partners may lack the necessary support from others to manage at home, support that they are often appreciative of when it is available [37].

Care partners and decision-making at end of life

The importance of advance care planning and continual informational support becomes more evident as the family care partner takes on the role of proxy decision-maker as dementia progresses. As discussed in previous chapters, for many care partners being a proxy decision-maker about end-of-life decisions follows experience making decisions during transitions such as hospital admissions or moves to assisted living or long-term care homes. In the long-term care home setting, family care partners tend to consider religious faith, quality of life, the provision of dignity and comfort, as well as their own personal experiences and advanced statements made by the person with dementia in their decisions. However, research demonstrates that for people with advanced dementia receiving care in a long-term care home, most proxy care preferences were for comfort-focused care (62.2%) care over basic (31.1%) or intensive medical care (6.5%) [53]. A preference for intensive medical care includes all available medical treatments. The person with advanced dementia may be admitted to the hospital or intensive care unit and receive treatments such as cardio pulmonary resuscitation, tube feeding, or mechanical ventilation. A preference for basic medical care does not involve intensive treatments listed previously, but the person with advanced dementia may receive some medical treatments, such as intravenous fluids or medications, and may be admitted to hospital for care. A preference for comfort-focused care means the person receives medical treatment to alleviate discomforting symptoms such as pain or shortness of breath, ideally "at home," and hospitalization is avoided [53].

Family care partners may require informational and emotional support in dealing with lost opportunities to elicit end-of-life care preferences from their loved one or with the clinical outcomes of a palliative approach to comfort care. Dempsey et al. [26] found that caregivers expressed regret for missing the opportunity to discuss death, while their loved one was able share their preferences. Family care partners may feel they are not in a position to engage in the discussion or simply do not recognize that they may lose the ability to have these conversations. Support is needed to help prepare family care partners for this role.

> *I think what's so important is that you make the most of the time you have. Why didn't I know mum would lose her speech [crying]? Because I should have spoken to her. I should have talked to her about death* [26] (pg. 13).

Negative feelings can be lessened when previous discussions and decisions have taken place [37,51]. However, family care partners can struggle when there is eventually a decision that provides comfort-focused care rather than active treatment. They can feel responsible for their loved one's death when active treatment such as intravenous fluid is not initiated or is withdrawn [11,26]. Indeed, goals of comfort-focused care may be ignored:

> *I don't know whether they cracked her ribs and I was just standing there calmly watching them and saying to my sister 'there's still a chance.' But it never occurred to me to ask them to stop because we did agree on DNR [do not resuscitate] but in the heat of that moment when you call 999 and when they ask you to do CPR you just do it* [51] (pg. 8).

These experiences highlight the importance of ongoing education and support for family care partners as they engage on decision-making, thus helping to facilitate the transition to end-of-life care.

The bereaved family care partner

In keeping with a palliative approach to care, when death of a person with dementia is imminent or they have very recently died, family care partners should be offered grief and bereavement support [2]. As noted by Durepos et al. [54], clinical guidelines for dementia care often lack direction on how formal care partners can help families prepare for and cope with the death of their relative affected by dementia. Importantly, formal care partners should recognize that family's feelings of loss and grief for their relative may start early in the individual's progression of the disease [37], meaning that support should not be limited to when loss is imminently expected. In addition, family members may feel guilt from their relative's end-of-life care experience [51], as well as have difficulty in adjusting to life after relinquishing caring responsibilities and resuming their role in society [26] ... "*you have the time but you just don't want to do anything*" (pg. 15).

Optimal assessment and treatment of symptoms

A hallmark of the palliative approach to care is the attention to, and treatment of, pain and other symptoms that may arise over the course of a person's illness [16]. For people with dementia, special consideration is required both for how symptoms are assessed in the evolving nature of the illness and what symptoms may be most prevalent as they approach the end of life. As symptoms are highly subjective and individually experienced, self-report is considered the gold standard for assessment [55]. Care providers have noted success in using self-report or simple visual tools such as the faces scale in people with dementia. Therefore the use of proxy reports or observational tools should be limited to individuals who are no longer able to reliably articulate their experience [56].

Pain

A consistent challenge documented for those with dementia is the underassessment and undertreatment of pain, often owing to the fact that in advanced stages of dementia, people cannot verbally express that they are experiencing it [57]. This is troubling in light of a recent study illustrating that approximately 60% of nursing home residents with cognitive impairment experience either consistent low or mild pain in their last 6 months of life, and about 34% experience either high or increasing pain levels during this time [58]. This finding of persistent pain over consecutive, regular assessments, and that pain increases significantly over the last weeks of life, has been documented by others [59–61]. It is vital that care providers recognize that for people with dementia, pain perceptions may remain intact, and cognitive changes may block the memory and interpretation of pain but not the experience of it [62]. Healthcare providers may recognize painful experiences ...

> *her muscles had just deteriorated and there was no substance to her body ... When she was sitting, it was like she was sitting on her hipbones rather than any cushioning ... most of her pain was sitting* [36] (pg. 32).

As such, assessment of pain in people with dementia must become focused on the behaviors that may be indicative of pain, not on its verbal expression. Accurate assessment forms the basis of pain treatment [63], including proxy behavioral observation [64]. The foundation of many of the observational tools is on how the pain is manifested, such as through facial expressions, negative vocalizations, body language, changes in activity patterns, or changes in interpersonal interactions [55,65]. A number of tools are available for the assessment of pain in people with dementia, but in their recent meta-review, Lichtner et al. [55] concluded that no one tool could be recommended because of limited evidence regarding reliability, validity, and clinical utility. It has, therefore, been suggested that healthcare providers use a combination of approaches to determine pain experiences, including taking into account reports from the individuals themselves, their care partners, past medical history and pain history including current comorbidities and painful conditions, and interpretation of changes in mood, behavior, and other nonverbal cues [66].

Symptom burden

For people with dementia, there are a number of additional physical symptoms that can contribute to the discomfort and distress of the individual as they transition into the end stage of their illness. These symptoms typically include agitation and delirium, dysphagia, shortness of breath, and infections such as pneumonia, septicemia, and urinary tract infections [67]. There is a wide variation in the prevalence of these symptoms among people with dementia. For example, shortness of breath has been reported in 8%–80% of individuals [59,68] and agitation in 35%–71% near the end of life [59,69]. When examining the symptom experience of people with dementia in their last months of life, nearly half of the participants in a study by Mitchell et al. [28] experienced an infection and 86% developed eating challenges; survival was poor when these symptoms developed. These findings have been articulated by others [67]. The management of these symptoms is key to achieving a good death experience and dying peacefully, from the perspective of family care partners. Lower levels of physical and emotional distress are associated with positive assessments of dying peacefully [70] and are typically hallmarks of a good death reported by health care providers and care partners [2].

What is important to understand with respect to people with dementia is that these symptoms may be interrelated and the concept of total pain may be useful in guiding a holistic approach to pain and symptom management [71,72]. The idea of total pain recognizes that symptoms do not occur in isolation and may greatly influence each other. Pain may be associated with and triggered by agitation or depression; shortness of breath may arise from anxiety. It also expands upon the notion that pain arises only from physical causes. Total pain may manifest from spiritual, psychological, and/or social causes. As such, healthcare providers must expand traditional notions of what it means to experience pain and to explore a range of possibilities in order to provide a positive dying experience and to reduce symptom distress of people with dementia.

Beyond physical care psychosocial and spiritual support

In addition to tending to physical care needs, spirituality has always been regarded as a central component of a holistic approach to palliative care [73]. Although once understood in a strictly religious context, it is now often considered to reflect vital principles of existence and well-being more generally. A consensus definition for spiritual care remains elusive [73,74]. Spirituality is a complex construct involving personal dimensions around meaning or satisfaction with life, relationships and connectedness, and/or transcendence [75]. Despite its centrality, spirituality has not received the same research attention as the other bio-psycho-social palliative care components [73,75]. Indeed, spiritual care considerations are seldom included in clinical dementia care guidelines [54].

The need for good advance care planning discussions, including expression of values and beliefs, becomes evident as healthcare providers help to support the spiritual needs of people with dementia as they transition to end of life. In recognizing the importance of spirituality in dementia palliative care, van der Steen et al. [2] recommend an assessment of religious affiliation and involvement, as well as offering religious activities and other sources of spiritual support. However, loss of a person's ability to verbally communicate during this time means that a spiritual care assessment will be limited to the family care partner's experiences, the individual's advanced statements, and the individual's responses to spiritual care interventions.

Impaired ability to communicate does not preclude people with advanced dementia from experiencing distress about death and dying. However, care partners and healthcare providers may assume that spiritual care is no longer relevant in advanced stages of dementia. However, implicit memory and habits of religious and spiritual practices mean that people with dementia respond to music and symbols associated with those practices. Scott [76] suggests engaging in spiritual activities with "reduced cognitive demand and high emotional content … [such as] holding a religious object or book, singing familiar hymns or sacred songs, spinning a prayer wheel, reciting favorite prayers or popular scriptures, being helped to face Mecca to pray, performing simple yoga exercises or chanting" (p. 47). Similarly, Toivonen et al. [77] found that nurses could learn about and support the personal philosophies of older adults with dementia through prayer, discussing spiritual topics (e.g., existence of God or the meaning of life and death), spiritual reminiscence (e.g., through familiar spiritual music, objects or pictures), reading or listening to spiritual literature, listening to spiritual music, and supporting connectedness to members of the religious community. The engagement of individuals in rituals and the use of music have been noted to have significant positive effects on those with advanced dementia [78].

Family care partners recognize the importance of treating their relative with respect and dignity as they progress to the later stages of the disease. While family care partners may not explicitly think about this as meeting spiritual needs, they recognize their relatives' needs for personal connectedness and meaning that are elements of spirituality [38]. Often, it is connection to nature, to one's god, or with others, that remain vital to a person's spirituality [79]. With respect to individuals with advanced dementia and severe impairment, it is important for health care providers to promote their quality of living, rather than mere "existing," with meaningful activities and supports for engagement [80]. As described by the care partner of person with dementia *"[w]hen asked if she ever became bedfast, he said no. 'I kept getting her up in the wheelchair every day so we could feel we were together and not just me sitting by the bed.'"* [36] (pg. 32). Ennis and Kazer [78] identify the value of presence, showing kindness, and establishing trusting environments as significant spiritual nursing interventions. By promoting quality of living through these varied means, attending to the spiritual needs of individuals may be met.

Transition to end-of-life and palliative care—final considerations

Supporting people with dementia and their family care partners to achieve a good-quality end of life and death is underpinned by a palliative approach to care [2]. To date, research examining the lived experiences of people and their care partners as they navigate end-of-life care highlights the need for continued effort in providing holistic palliative care, tailored to individual and family care partner needs. This chapter draws attention to some key areas to address in supporting a palliative approach, including symptom management, advance care planning, family involvement, and spiritual care. However, perhaps the greatest challenge to overcome in advancing a palliative approach to dementia care is achieving widespread understanding of the terminal nature of the disease, resulting from severe cognitive and functional impairments [28]. A common appreciation of the terminal nature of dementia could have substantial implications:

(1) Initiation of a palliative approach to dementia care at the point of diagnosis,

(2) Recognition of the need for ongoing dementia care education targeted at people with dementia as well as healthcare providers and family care partners to support goal-directed decision-making,

(3) Early initiation of holistic advance care planning, where people with dementia and their healthcare providers and family care partners share their goals, values, and preferences in advance of crisis, and

(4) Better preparation of family care partners for care provision and support for the people with dementia, as well as grief and bereavement support from healthcare providers.

The voices of people with dementia, their care partners, and healthcare providers need to be continually sought and shared to help in achieving high-quality care at end of life. It is through understanding their experiences that the palliative approach to care for people with dementia may be best be shaped to meet their needs.

References

[1] Iaboni A, Van Ooteghem K, Flint AJ, Keren R, Grossman D. The value of a palliative frame in advanced dementia. Am J Geriatr Psychiatry 2018;26(8):906–7. Available from: https://ac-els-cdn-com.uml.idm.oclc.org/S1064748118303221/1-s2.0-S1064748118303221-main.pdf?_tid=9e13a73c-3c6d-4f3b-9473-ac89ba140e9b&acdnat=1541622236_0322c79301c1dd88ebf6708f19289758.

[2] van der Steen JT, Radbruch L, Hertogh C, De Boer ME, Hughes JC, Larkin P, et al. White paper defining optimal palliative care in older people with dementia: a Delphi study and recommendations from the European Association for Palliative Care. Palliat Med 2014;28(3):197–209.

[3] Palan Lopez R, Mitchell SL, Givens JL. Preventing burdensome transitions of nursing home residents with advanced dementia: it's more than advance directives. J Palliat Med 2017;20(11):1205–9. Available from: www.liebertpub.com.

[4] Erel M, Marcus E-L, Dekeyser-Ganz F. Barriers to palliative care for advanced dementia: a scoping review. Ann Palliat Med 2017;6(4):365–79. Available from: http://apm.amegroups.com/article/view/15568/15611.

[5] Van Der Steen JT, Onwuteaka-Philipsen BD, Knol DL, Ribbe MW, Deliens L. Caregivers' understanding of dementia predicts patients' comfort at death: a prospective observational study. BMC Med 2013;11:105. Available from: http://www.biomedcentral.com/1741-7015/11/105.

[6] Ryan T, Gardiner C, Bellamy G, Gott M, Ingleton C. Barriers and facilitators to the receipt of palliative care for people with dementia: the views of medical and nursing staff. Palliat Med 2011;26(7):879–86. Available from: https://journals-sagepub-com.uml.idm.oclc.org/doi/pdf/10.1177/0269216311423443.

[7] Forbes S, Bern-Klug M, Gessert C. End-of-life decision making for nursing home residents with dementia. J Nurs Scholarsh 2000;32(3):251–8.

[8] Gessert C, Forbes S, Bern-Klug M. Planning end-of-life care for patients with dementia: roles of families and health professionals. Omega J Death Dying 2001;42(4):273–91.

[9] Palan Lopez R. Doing what's best: decisions by families of acutely ill nursing home residents. West J Nurs Res 2009;31:613–26. Available from: http://wjn.sagepub.comhttp//online.sagepub.com.

[10] Andrews S, Mcinerney F, Toye C, Parkinson C-A, Robinson A. Knowledge of dementia: do family members understand dementia as a terminal condition? Dementia 2017;16(5):556–75. Available from: https://journals.sagepub.com/doi/pdf/10.1177/1471301215605630.
[11] Hennings J, Froggatt K, Keady J. Approaching the end of life and dying with dementia in care homes: the accounts of family carers. Rev Clin Gerontol May 22, 2010;20(02):114–27. Available from: http://www.journals.cambridge.org/abstract_S0959259810000092.
[12] Lunney JR, Lynn J, Foley DJ, Lipson S, Guralnik JM. Patterns of functional decline at the end of life. J Am Med Assoc 2003;289(0098–7484; 18):2387–92.
[13] Bartley MM, Suarez L, Reem, Shafi MA, Baruth JM, Benarroch AJM, et al. Dementia care at end of life: current approaches. Curr Psychiatr Rep 2018;20:50. Available from: https://doi.org/10.1007/s11920-018-0915-x.
[14] Hanson E, Hellströ A, Sa Sandvide Å, Jackson GA, Macrae R, Waugh A, et al. The extended palliative phase of dementia: an integrative literature review. Dementia 2016;0(0):1–27. Available from: http://www.uws.ac.uk/palliareproject/.
[15] Mitchell SL, Kiely DK, Hamel MB, Park PS, Morris JN, Fries BE. Estimating prognosis for nursing home residents with advanced dementia. J Am Med Assoc June 2004;291(22):2734–40.
[16] Kelley AS, Morrison RS. Palliative care for the seriously ill. N Engl J Med 2015;373(3):747–55. Available from: doi: 0.1056/NEJMra1404684.
[17] Brown MA, Sampson EL, Jones L, Barron AM. Prognostic indicators of 6-month mortality in elderly people with advanced dementia: a systematic review. Palliat Med May 22, 2013;27(5):389–400. Available from: http://journals.sagepub.com/doi/10.1177/0269216312465649.
[18] Van Der Steen JT, Radbruch L, De Boer ME, Jünger S, Hughes JC, Larkin P, et al. Achieving consensus and controversy around applicability of palliative care to dementia. Int Psychogeriatrics C Int Psychogeriatr Assoc 2016;28(1):133–45.
[19] Birch D, Draper J. A critical literature review exploring the challenges of delivering effective palliative care to older people with dementia. J Clin Nurs May 1, 2008;17(9):1144–63. Available from: http://doi.wiley.com/10.1111/j.1365-2702.2007.02220.x.
[20] van Riet Paap J, Mariani E, Chattat R, Koopmans R, Kerhervé H, Leppert W, et al. Identification of the palliative phase in people with dementia: a variety of opinions between healthcare professionals. BMC Palliat Care December 4, 2015;14(1):56. Available from: http://bmcpalliatcare.biomedcentral.com/articles/10.1186/s12904-015-0053-8.
[21] Robinson A, Eccleston C, Annear M, Elliott K, Andrews S, Stirling C, et al. Who knows, who care? Dementia knowledge among nurses, care workers, and family member of people living with dementia. J Palliat Care 2014;30(3):158–65. Available from: https://search-proquest-com.uml.idm.oclc.org/docview/1584944624/fulltextPDF/2824C97671C241BEPQ/1?accountid=14569.
[22] Covinsky KE, Eng C, Lui LY, Sands LP, Yaffe K. The last 2 years of life: functional trajectories of frail older people. J Am Geriatr Soc January 2003;51(4):492–8. Available from: http://www.blackwell-synergy.com/links/doi/10.1046/j.1532-5415.2003.51157.x/abs.
[23] Cox S, Cook A. Caring for people with dementia at the end of life. In: Hockley J, Clarke D, editors. Palliative care for older people in care homes. Buckingham: Open University Press; 2002. p. 86–103. Buckingham.
[24] Hill SR, Mason H, Poole M, Vale L, Robinson L. What is important at the end of life for people with dementia? The views of people with dementia and their carers. Int J Geriatr Psychiatry September 1, 2017;32(9):1037–45. Available from: http://doi.wiley.com/10.1002/gps.4564.

[25] Shanley C, Russell C, Middleton H, Simpson-Young V. Living through end-stage dementia: the experiences and expressed needs of family carers. Dementia August 2011;10(3):325–40. Available from: http://search.ebscohost.com/login.aspx?direct=true&db=rzh&AN=104663609&site=ehost-live.
[26] Dempsey L, Dowling M, Larkin P, Murphy K. Providing care for a person with late-stage dementia at home: what are carers' experiences? Dementia 2018;0(0).
[27] Mitchell SL, Kiely DK, Hamel MB. Dying with advanced dementia in the nursing home. Arch Intern Med February 2004;164(3):321–6.
[28] Mitchell SL, Teno JM, Kiely DK, Shaffer ML, Jones RND, Prigerson HG, et al. The clinical course of advanced dementia. N Engl J Med 2009;361(16):1529–38. Available from: https://www-nejm-org.uml.idm.oclc.org/doi/pdf/10.1056/NEJMoa0902234.
[29] Carter G, McLaughlin D, Kernohan WG, Hudson P, Clarke M, Froggatt K, et al. The experiences and preparedness of family carers for best interest decision-making of a relative living with advanced dementia: a qualitative study. J Adv Nurs July 2018;74(7):1595–604. Available from: http://search.ebscohost.com/login.aspx?direct=true&db=rzh&AN=130452284&site=ehost-live.
[30] World Health Organization. Dementia a public health priority [Internet]. 2012. Available from: http://apps.who.int/iris/bitstream/handle/10665/75263/9789241564458_eng.pdf;jsessionid=CCBF985EC184DE56550F9ECEEB3B1DB5?sequence=1.
[31] Sawatzky R, Porterfield P, Roberts D, Lee J, Liang L, Reimer-Kirkham S, et al. Embedding a palliative approach in nursing care delivery an integrated knowledge synthesis. Adv Nurs Sci 2017;40(3):263–79.
[32] Hines S, McCrow J, Abbey J, Foottit J, Wilson J, Franklin S, et al. The effectiveness and appropriateness of a palliative approach to care for people with advanced dementia: a systematic review. JBI Libr Syst Rev 2011;9(26):960–1131. Available from: http://www.epistemonikos.org/documents/258a347b2136c94298cfb250a0b1a442d56dd1a6.
[33] Kristjanson L, Toye C, Dawson S. New dimensions in palliative care: a palliative approach to neurodegenerative diseases and final illness in older people. Med J Aust 2003;179(6Suppl. l):S41–3. Available from: https://www-mja-com-au.uml.idm.oclc.org/journal/2003/179/6/new-dimensions-palliative-care-palliative-approach-neurodegenerative-diseases.
[34] Sawatzky R, Porterfield P, Lee J, Dixon D, Lounsbury K, Pesut B, et al. Conceptual foundations of a palliative approach: a knowledge synthesis. BMC Palliat Care December 2016;15(1):5.
[35] Thompson GN, Roger K. Understanding the needs of family caregivers of older adults dying with dementia. Palliat Support Care June 18, 2014;12(03):223–31. Available from: http://www.journals.cambridge.org/abstract_S1478951513000461.
[36] Glass AP. Family caregiving and the site of care: four narratives about end-of-life care for individuals with dementia. J Soc Work End Life Palliat Care 2016;12(1–2):23–46. Available from: http://www.tandfonline.com/action/journalInformation?journalCode=wswe20.
[37] Broady TR, Saich F, Hinton T. Caring for a family member or friend with dementia at the end of life: a scoping review and implications for palliative care practice. Palliat Med 2018;32(3):643–56. Available from: https://doi.org/10.1177/0269216317748844.
[38] Davies N, Rait G, Maio L, Iliffe S. Family caregivers' conceptualisation of quality end-of-life care for people with dementia: a qualitative study. Palliat Med 2017;31(8):726–33. Available from: https://journals-sagepub-com.uml.idm.oclc.org/doi/pdf/10.1177/0269216316673552.
[39] Zahradnik EK, Grossman H. Palliative care as a primary therapeutic approach in advanced dementia: a narrative review. Clin Ther 2014;36(11):1512–7. Available from: https://doi.org/10.1016/j.clinthera.2014.10.006.

[40] van der Steen JT. Dying with dementia: what we know after more than a decade of research. J Alzheimer's Dis 2010;22:37–55. Available from: https://content-iospress-com.uml.idm.oclc.org/download/journal-of-alzheimers-disease/jad100744?id=journal-of-alzheimers-disease/jad100744.
[41] Sudore RL, Lum HD, You JJ, Hanson LC, Meier DE, Pantilat SZ, et al. Defining advance care planning for adults: a consensus definition from a multidisciplinary Delphi panel. J Pain Symptom Manag 2017;53(5):821–32. e1. Available from: https://doi.org/10.1016/j.jpainsymman.2016.12.331.
[42] Mitchell SL. Care of patients with advanced dementia [Internet]. 2018. Available from: https://www.uptodate.com/contents/care-of-patients-with-advanced-dementia.
[43] Dixon J, Karagiannidou M, Knapp M. The effectiveness of advance care planning in improving end-of-life outcomes for people with dementia and their carers: a systematic review and critical discussion. J Pain Symptom Manag 2018;55(1):132–150.e1.
[44] Piers R, Albers G, Gilissen J, De Lepeleire J, Steyaert J, Van Mechelen W, et al. Advance care planning in dementia: recommendations for healthcare professionals. BMC Palliat Care 2018;17(1):1–17.
[45] Ashton SE, Roe B, Jack B, McClelland B. End of life care: the experiences of advance care planning amongst family caregivers of people with advanced dementia – a qualitative study. Dementia September 2016;15(5):958–75. Available from: http://search.ebscohost.com/login.aspx?direct=true&db=rzh&AN=117960724&site=ehost-live.
[46] McCleary L, Thompson GN, Venturato L, Wickson-Griffiths A, Hunter P, Sussman T, et al. Meaningful connections in dementia end of life care in long term care homes. BMC Psychiatry December 24, 2018;18(1):307. Available from: https://bmcpsychiatry.biomedcentral.com/articles/10.1186/s12888-018-1882-9.
[47] Robinson L, Dickinson C, Bamford C, Clark A, Hughes J, Exley C. A qualitative study: professionals' experiences of advance care planning in dementia and palliative care, "a good idea in theory but···. Palliat Med 2013;27(5):401–8. Available from: https://doi.org/10.1177/0269216312465651.
[48] Mogan C, Lloyd-Williams M, Dening KH, Dowrick C. The facilitators and challenges of dying at home with dementia: a narrative synthesis. Palliat Med 2018;32(6):1042–54. Available from: https://doi.org/10.1177/0269216318760442.
[49] Kaasalainen S, Sussman T, Durepos P, Mccleary L, Ploeg J, Thompson G. What are staff perceptions about their current use of emergency departments for long-term care residents at end of life? Clin Nurs Res 2017;00:1–16. Available from: https://journals-sagepub-com.uml.idm.oclc.org/doi/pdf/10.1177/1054773817749125.
[50] Thompson GN, McClement SE, Menec VH, Chochinov HM. Understanding bereaved family members' dissatisfaction with end-of-life care in nursing homes. J Gerontol Nurs October 1, 2012;38(10):49–60. Available from: http://www.healio.com/doiresolver?doi=10.3928/00989134-20120906-94.
[51] Moore KJ, Davis S, Gola A, Harrington J, Kupeli N, Vickerstaff V, et al. Experiences of end of life amongst family carers of people with advanced dementia: longitudinal cohort study with mixed methods. BMC Geriatr 2017;17:135. Available from: https://bmcgeriatr.biomedcentral.com/track/pdf/10.1186/s12877-017-0523-3.
[52] Jones K, Birchley G, Huxtable R, Clare L, Walter T, Dixon J. End of life care: a scoping review of experiences of advance care planning for people with dementia. Dementia 2019;18(3):825–45. Available from: https://journals-sagepub-com.uml.idm.oclc.org/doi/pdf/10.1177/1471301216676121.
[53] Mitchell SL, Palmer JA, Volandes AE, Hanson LC, Habtemariam D, Shaffer ML. Level of care preferences among nursing home residents with advanced dementia. J Pain Symptom Manag 2017;54(3):340–5.

[54] Durepos P, Wickson-Griffiths A, Hazzan AA, Kaasalainen S, Vastis V, Battistella L, et al. Assessing palliative care content in dementia care guidelines: a systematic review. J Pain Symptom Manag 2017;53(4):804–13.
[55] Lichtner V, Dowding D, Esterhuizen P, Closs SJ, Long AF, Corbett A, et al. Pain assessment for people with dementia: a systematic review of systematic reviews of pain assessment tools. BMC Geriatr December 17, 2014;14(1):138. Available from: http://bmcgeriatr.biomedcentral.com/articles/10.1186/1471-2318-14-138.
[56] Pautex S, Michon A, Guedira M, Emond H, Le Lous P, Samaras D, et al. Pain in severe dementia: self-assessment or observational scales? J Am Geriatr Soc July 1, 2006;54(7):1040–5. Available from: http://doi.wiley.com/10.1111/j.1532-5415.2006.00766.x.
[57] Yeun-Sim Jeong S, Tsai I-P, Sarah Yeun-Sim Jeong C, Hunter S. Pain assessment and management for older patients with dementia in hospitals: an integrative literature review. Pain Manag Nurs 2018;19(1):54–71. Available from: https://doi.org/10.1016/.
[58] Thompson GN, Doupe M, Colin R, Baumbusch J, Estabrooks CA. Pain trajectories of nursing home residents nearing death. JAMDA 2017;18. 700–6. Available from: https://doi.org/10.1016/j.jamda.2017.03.002.
[59] Hendriks SA, Smalbrugge M, Galindo-Garre F, Hertogh CM, van der Steen JT. From admission to death: prevalence and course of pain, agitation, and shortness of breath, and treatment of these symptoms in nursing home residents with dementia. J Am Med Dir Assoc June 2015;16(6):475–81. Available from: https://doi.org/10.1016/j.jamda.2014.12.016.
[60] Koppitz A, Bosshard G, Schuster DH, Hediger H, Imhof L. Type and course of symptoms demonstrated in the terminal and dying phases by people with dementia in nursing homes. Z Gerontol Geriatr 2015;48:176–83. Available from: https://link-springer-com.uml.idm.oclc.org/content/pdf/10.1007/s00391-014-0668-z.pdf.
[61] Estabrooks CA, Hoben M, Poss JW, Chamberlain SA, Thompson GN, Silvius JL, et al. Dying in a Nursing Home: Treatable Symptom Burden and its Link to Modifiable Features of Work Context. J Am Med Dir Assoc. June 1; 16(6); 515–520. March 2015. https://doi.org/10.1016/j.jamda.2015.02.007.
[62] Olson E. Dementia and neurodegenerative diseases. In: Morrison RS, Meier DE, editors. Geriatric palliative care. New York, New York: Oxford University Press; 2003. p. 160–72.
[63] Fink RM, Gates RA, Montgomery RK. Pain assessment. In: Ferrell BR, Coyle N, Paice JA, editors. Oxford textbook of palliative nursing. 4th ed. New York, New York: Oxford University Press; 2015. p. 113–34.
[64] Burns M, McIlfatrick S. Palliative care in dementia: literature review of nurses' knowledge and attitudes towards pain assessment. Int J Palliat Nurs August 2, 2015;21(8):400–7. Available from: http://www.magonlinelibrary.com/doi/10.12968/ijpn.2015.21.8.400.
[65] Strand LI, Gundrosen KF, Lein RK, Laekeman M, Lobbezoo F, Defrin R, et al. Body movements as pain indicators in older people with cognitive impairment: a systematic review. Eur J Pain 2019, April;23(4):669–85. Available from: https://doi.org/10.1002/ejp.1344.
[66] Jansen BDW, Brazil K, Passmore P, Buchanan H, Maxwell D, Mcilfatrick SJ, et al. "There's a Catch-22'-The complexities of pain management for people with advanced dementia nearing the end of life: a qualitative exploration of physicians" perspectives. Palliat Med 2017;31(8):734–42. Available from: https://doi.org/10.1177/0269216316673549.

[67] Sampson EL, Candy B, Davis S, Gola AB, Harrington J, King M, et al. Living and dying with advanced dementia: a prospective cohort study of symptoms, service use and care at the end of life. Palliat Med 2018;32(3):668–81. Available from: https://doi.org/10.1177/0269216317726443.
[68] Soares LGL, Japiassu AM, Gomes LC, Pereira R, Peçanha C, Goldgaber T. Prevalence and intensity of dyspnea, pain, and agitation among people dying with late stage dementia compared with people dying with advanced cancer: a single-center preliminary study in Brazil. Ann Palliat Med 2018;7(4):437–43. Available from: http://apm.amegroups.com/article/view/19619/19602.
[69] Hendriks SA, Smalbrugge M, Hertogh CMPM, Van Der Steen JT. Dying with dementia: symptoms, treatment, and quality of life in the last week of life. J Pain Symptom Manag 2014;47(4):710–20. Available from: https://doi.org/10.1016/j.jpainsymman.2013.05.015.
[70] De Roo ML, Albers G, Deliens L, De Vet HCW, Francke AL, Van N, et al. Physical and psychological distress are related to dying peacefully in residents with dementia in long-term care facilities. J Pain Symptom Manag 2015;50(1):1–8. Available from: https://doi.org/10.1016/j.jpainsymman.2015.02.024.
[71] Greenstreet W. The concept of total pain: a focused patient care study. Br J Nurs 2001;10(19):1248–55.
[72] Mehta A, Chan LS. Understanding of the concept of "Total Pain": a prerequisite for pain control. J Hospice Palliat Nurs 2008;10(1):26–32. Available from: http://downloads.lww.com/wolterskluwer_vitalstream_com/journal_library/njh_15222179_2008_10_1_26.pdf.
[73] Steinhauser K, Fitchett G, Handzo G, Johnson K, Koenig H, Pargament K, et al. State of the science of spirituality and palliative care pat I: definitions, measurement, and outcomes. J Pain Symptom Manag 2017;54(3):428–40. Available from: http://www.tandfonline.com/doi/full/10.1271/bbb.65.2205.
[74] Wein S. Spirituality — the psyche or the soul? Palliat Support Care 2014;12(02):91–4. Available from: http://www.journals.cambridge.org/abstract_S1478951514000303.
[75] Puchalski CM. Spirituality and the care of patients at the end-of-life : an essential component of care. Omega 2008;56(1):33–46.
[76] Scott H. The importance of spirituality for people living with dementia. Nurs Stand February 17, 2016;30(25):41–50. Available from: http://journals.rcni.com/doi/10.7748/ns.30.25.41.s47.
[77] Toivonen K, Charalambous A, Suhonen R. Supporting spirituality in the care of older people living with dementia: a hermeneutic phenomenological inquiry into nurses' experiences. Scand J Caring Sci 2018;32(2):880–8.
[78] Ennis EM, Kazer MW. The role of spiritual nursing interventions on improved outcomes in older adults with dementia. Holist Nurs Pract March 1, 2013;27(2):106–13. Available from: http://www.ncbi.nlm.nih.gov/pubmed/23399710.
[79] Toivonen K, Stolt M, Suhonen R. Nursing support of the spiritual needs of older adults living with dementia: a narrative literature review. Holist Nurs Pract September 1, 2015;29(5):303–12. Available from: http://www.ncbi.nlm.nih.gov/pubmed/26263290.
[80] Volicer L, Simard J. Palliative care and quality of life for people with dementia: medical and psychosocial interventions. Int Psychogeriatr 2015;27(10):1623–34.

CHAPTER 10

Steering through the waves and adjusting to transitions in dementia

Linda Garcia[1,2] Neil Drummond[3], Lynn McCleary[4]
[1]Life Research Institute, University of Ottawa, Ottawa, ON, Canada; [2]Interdisciplinary School of Health Sciences, Faculty of Health Sciences, University of Ottawa, Ottawa, ON, Canada; [3]Alberta Health Services Chair in Primary Care Research, Department of Family Medicine, University of Alberta, Canada; [4]Department of Nursing, Brock University, St. Catharines, Ontario, Canada

Introduction

As is the case in many chronic diseases, nobody expects to experience the impact of dementia. When people recognize that this will actually be one of the life challenges they face, few are fully prepared. As with many chronic diseases, the trajectory of dementia with respect to which transitions will be encountered is not always predictable, but it is clear that the experience of progressive decline, perhaps to the extent of eventual death, will result in a series of transitions of varying intensity and duration [1].

In this book we chose to use the metaphor of paddling through white water to represent navigating the numerous transitions people with dementia and their carers must maneuver. The experience of dementia is different for everyone; but as with the crests and troughs of the waves for a canoeist, dementia will challenge us both physically and mentally. The key to riding the waves of the dementia journey lies in our ability to know what is coming, to use the information we have about how to manage the waves, and to know when to ride them out and when to seek help. Wilson et al. [2] found that injuries and death were more common in those who canoe or kayak white waters rather than among those who engage in white water rafting. One of their explanations is that rafting often involves professionals who guide and help people, whereas paddlers are more likely to try to manage the waters on their own. Other factors are that canoes and kayaks are small and fairly unstable craft crewed by individuals or pairs, while rafts are large, stable, and almost always crewed by several people at once. Most people with dementia and their carers tend to be rather more like canoeists than rafters. They are pretty much on their own. Hence the chapters in this

Evidence-informed Approaches for Managing Dementia Transitions
ISBN 978-0-12-817566-8
https://doi.org/10.1016/B978-0-12-817566-8.00010-3

book have highlighted the need to seek help and information. Managing the ebb and flow of the experiences of dementia cannot easily be done alone. The authors of the chapters have proposed some common messages for people living with dementia and those who are part of their circle of care, either professionally or not. The purpose of this final chapter is to summarize these messages so that service providers might enhance their approaches in fostering the well-being of persons with dementia and family carers experiencing transitions.

Dementia is "typically defined as a clinical syndrome of cognitive decline that is sufficiently severe to interfere with social or occupational functioning." (page 2) [3]. Individuals with dementia often recognize the functional changes that are affected by memory loss, judgment, thinking skills, visual perception problems, or their ability to follow conversations. In some cases, these signs and symptoms are attributed to normal aging and individuals with dementia do not consult professionals. However, with more public awareness of the characteristics of dementia, individuals and their families are more inclined to recognize which symptoms are indicative of dementia. In many cases, the motivation to seek professional advice stems from the realization that the person's functioning is not what it was before the onset of symptoms.

If and when a diagnosis of dementia is obtained, a series of typical and highly likely transitions will occur. In some cases, some of these transitions have occurred before the diagnosis. Drummond et al. [4] reported on the perceived quality of nine such transitions:

1. initial problem identification
2. requiring support from extrafamilial sources
3. driving cessation
4. loss of financial autonomy
5. acute hospitalization
6. change in carer
7. relocation to a new community-based accommodation
8. relocation to long-term residential care and
9. entry to palliative and end-of-life care.

Not all persons with dementia will go through each transition, but many are common. As described in Chapter 1, much of the existing literature has focused on the experiences of people with dementia and their loved ones as they navigate any given transition, but we were unable to find any work that discusses the commonalities across these transitions. The authors of the preceding chapters have offered their interpretations of the evidence characterizing the experience of each transition identified by the Dementia Transition Study (Chapter 1). In each chapter, there is reference to persons with dementia and their families experiencing a critical juncture or decisive point where they are required to act, react, or adjust so that equilibrium can once again be attained. Whenever possible, it is desirable that this equilibrium results in some level of happiness or well-being.

Kopsov [5] recently presented the scientific community with a new model of subjective well-being which is well suited to our discussion about transitions and dementia. He formulates well-being not "as a stand-alone phenomenon but as a reflection and result of life progression" (p. 104). There is some merit in applying Kopsov's dynamic model to interpret what we have learned about dementia and transitions throughout this book. The dynamic process presented by his model is fueled by an assessment of one's needs, and the risks associated with meeting or not meeting those needs, using what knowledge and experiences we have at our disposal. In the end, satisfaction of these needs leads to subjective well-being, which he equates with happiness. He designed his model to be dynamic, where new knowledge and experiences help us readjust our risk assessments as well as the priorities of our needs. One can appreciate that this approach can help explain how one might navigate a transition. Faced with dementia, the cognitive and/or affective resources at our disposal will impact our assessments of risk and needs and it is therefore important that others help to fill those gaps to guide the person with dementia through the transition so that equilibrium and well-being can be restored.

Without delving too deeply into the complex model presented by Kopsov, an attempt will be made here to apply it to the experience of dementia. The person with dementia and carers might detect changes both inside and outside of themselves such as cognitive changes, affective changes, or changes in functioning. This will likely create a disequilibrium which will lead to the creation of needs and a potential transition. There have been ample discussions in this book about the importance of recognizing these changes and communicating with families and people living with dementia regarding what is known and available to them and whether they should consult.

Applying the Kopsov model, an individual with dementia and their carer might, based on the cognitive and affective resources available to them, past experiences and expectations, assess the likelihood that the newly created needs are likely to be satisfied. If they deem that this is unlikely, then the needs will go unmet and subjective well-being is unlikely. Carers, people with dementia, and service providers may all have different assessments of the likelihood of whether needs will be met. We have seen throughout this book that expectations play an important role in how people move forward or whether they move forward at all. For instance, lack of trust in the services offered will likely result in an appraisal that the need is not going to be met.

If the individual concludes that the need is likely to be met, then application of Kopsov's model would suggest that a risk analysis will take place as to whether the person should engage in an action. For instance, a person with dementia or their family may use prior experiences regarding driving to decide that some intervention regarding driving cessation is necessary, given the level of risk associated with cognitive impairment. If the risk is unacceptable, the person will likely engage in a smooth transition to driving cessation. As mentioned previously, carers, people with dementia, and service providers may all arrive at different risk assessments. Options can then be planned for future action. In the book, we have seen how engaging people with dementia and their families in the planning process will help in assessing appropriate options for action. Whether or not to move to an alternate level of living space will be based on knowledge and experiences of people with dementia and their families. Once acted upon, the results of these actions then serve as experiences that will further influence expectations and behavior around future needs. Trying to maintain well-being throughout the transitions will be dependent on a delicate balance of perceived need, risk, and likelihood of resolution. In the face of deteriorating cognition, this process will necessitate the involvement of several players to redress the balance after the disruption caused by the change and the need for the transition.

In this final chapter, we use the acronym R-E-A-D-J-U-S-T to help draw some overall, common messages from the preceding chapters. The acronym refers to recognition, expectations, autonomy, dementia-friendly environments, judicious changes, understanding and creating new aspects of self, savoring, and team of helpers. These strategies are offered in the hopes that individuals who live with dementia might be better served to navigate the bumpy waters of the dementia transitions.

Recognition

The first advice for navigating the waters is to be aware of the importance of recognizing the signs of change related to dementia, starting with the earliest traces of a pathway leading to a diagnosis, right through to those of imminent death. It is key that service providers consider the perspectives of not only the person with dementia but also their families and friends, who may notice some of these signs earlier than the professionals. All quotes are from the Dementia Transition Study described in Chapter 1.

Quote 1

Interviewer: I see. Um, so your advice would be to recognize that early signs of short term memory could be something more than normal aging?

Carer: Yeah, that's, that's not an easy, it's easy to say you'd watch for it, but unless there is something, sort of dramatic happens, ah, it may be very, very difficult to.

Quote 2

Interviewer: Is there a particular event that, that shouldn't happen in that way, that makes you think, "Uh-oh, something is wrong"? Or is that day to day regular small stuff? Or is there big stuff happening?

Carer: Well, no. No, no, small stuff leading up, you know, not being able to dial and such a thing. Before, but this, this just really, well affected him badly. He, just like I said, he couldn't figure out how to do things.

Delay in recognizing and learning about people's needs which may lead to transition (e.g., diagnosis, change in residence, preparing for end of life) usually means a delay in accessing services that may help. Carers are pivotal in discussions, even during the peridiagnostic period, about what services they are ready to access (Chapters 2). They are often the ones who see the first signs of dementia, and they are the ones who might recognize that driving is erratic or might notice that the individual with dementia has difficulty managing their finances. Therefore service providers are urged to listen to those who are bringing these changes to their attention.

In Chapter 2, attention is brought to how people with dementia are sometimes confronted with physicians who minimize the significance of their symptoms. Given that no two older individuals are the same and given that their cultural, social, and physical contexts are different, it is important to create spaces that allow people with dementia and their families to express themselves and advocate for their needs. Personal biases regarding culture or personality should not cloud service providers' judgments of what a person with dementia and their loved ones are saying. It is important that families' accounts of what has changed be considered and not be judged. Service providers must realize that recognition of functional differences starts with those who are closest to the situation.

As awareness of the signs and symptoms of dementia, as well as the possible disease trajectories become more well known, so too will people's ability to recognize when to consult a health professional. There has been a strong emphasis throughout the book on the need for information. Information is key to being able to navigate the next transition, and it is often at the early stage of diagnosis that this information is given to people with dementia and carers. Yet, as evident throughout the book, people need to have information provided to them in ways that make sense given their stage of illness and functioning, throughout the course of the condition. In Chapter 9 it is argued that even preparation for end of life can occur at these early stages. Recognizing the chronic and terminal nature of dementias, advance care planning and early information about palliative care and end of life could reduce crises as the person nears death.

It is difficult for families and carers of people with dementia to grapple with dealing with present dementia transitions while also keeping an eye out for future change. Grieving for losses and anticipating the difficult road ahead requires readjustment in terms of hopes and expectations. Helping carers through these experiences, through consultation and support, should be an objective for every service provider. Advice on how to manage the services and where to find community support can go a long way to giving families the tools they need to navigate and know when to consult. This might be as fundamental as helping them recognize that a transition is about to happen and managing expectations about it.

Expectations

The second significant message across previous chapters is the need to understand the hopes and expectations of people with dementia and their families about what is to come. Knowing what people already know or think they know is the best starting point for addressing the gaps in knowledge.

Quote 1

Person with dementia: I must admit that, I have been thinking in the last couple of weeks. What about down the road in the next couple of years. And I'm looking around the house and thinking, now I haven't done this and that, we didn't change this, it looks pretty grimy. And then I am thinking okay, maybe a couple of years down the road I am going to think that a seniors' residence of a suitable kind would be a good idea. I have a summer cottage and it is large. It is a 4-hour drive. I love it and it has been in the family for a 100 years. I am very fond of that. I have the responsibility of that and the house. It is probably a little bit more than I will want to contend with for the next umpteen years. So that maybe if I was in a seniors' residence or something smaller, it would be easier.

Interviewer: Okay, that's a good thought process.

Person with dementia: This is some new thinking that has come along maybe within the last couple of weeks.

Interviewer: Wow Okay.

Person with dementia: I will let this kind of thing percolate. I think what would I want to do with that, who would want this, then I just turn it off and I say, look, relax. It is a lovely neighbourhood.

Interviewer: No but it's good you're thinking, thinking for the future is good.

Person with dementia: So as I say if you let it percolate a little bit, it is not an entirely new wish.

Quote 2

Interviewer: What would you have done, put your husband on a list sooner? Would you have checked facilities out or kind of put things into place sooner? Would you have done things differently in hindsight?

Carer: No, I wouldn't have done it sooner because I didn't expect it to change that fast because he was fairly level for a while and then all of a sudden — A big drop.

Chapter 1 clarifies the distinction between hopes and expectations. Hopes are what we wish for, prefer, or desire, while expectations are what we view as probable. In the beginning phases of the dementia journey, depending on the person's knowledge and personal previous experience with dementia, the person's hopes and expectations might be high. For instance, one might hope that the person will not end up in a long-term care home and one might expect that their functional ability will be minimally affected. As information is obtained and life is experienced, expectations might be adjusted and so might be the actions and reactions to each transition. As seen in Chapter 7, while adjusting to the transition to living in an assisted living residence, family carers hoped that they would not have to experience the transition to a long-term care home but at the same time expected it would happen and were preparing for it.

As seen throughout the book, clear communication about what is likely to come, what services and support are available, and how to access them are key in helping families adjust. Chapter 6 highlights how families who were able to recognize and expect differences in their loved ones and did not expect the person to remain as he or she once was, adjusted better to the care partner role.

In Chapter 3, it is suggested that better tests to predict drivers' capabilities are needed. Such tests would better inform expectations of persons with dementia, families, and healthcare providers. Furthermore, we are reminded that cessation of driving is a normative change with advanced age and most people eventually transition from the driver's seat to the passenger's. We are called on to accept this, with the rationale that if we all expected to stop driving one day, we would develop alternate transportation policies and personal plans, which would increase resilience when faced with the transition, whether as a result of a chronic disease (such as dementia) or another reason. Adjusting expectations could avert a crisis. Similar observations were made in Chapter 4 with regards to loss of financial autonomy. The authors draw attention to a strategy for addressing potential loss of financial autonomy—older adults communicating with their adult children about their finances before there is a need to delegate control. While older adults would hope not to have to transfer financial decisions to their adult children, the probability of having to do so increases with advancing age. Financial counselors are well positioned to work alongside other service providers to help people address the expectation of an eventual change in an ability to manage finances.

Expectation of future events is also important in managing the transition of hospitalization. As most would agree, people expect person-centered approaches when we are hospitalized, but this expectation is not frequently realized [6], especially for people with dementia (Chapter 5). Families expect the best care for their loved ones, but, for example, healthcare providers in hospitals may judge that rehabilitation is not needed for people with dementia, regardless of how good their functional status is.

In the later stages of the dementia trajectory, people are potentially faced with relocation to a collective dwelling, be it an assisted living residence or a long-term care home. Hopes and expectations once again need to be adjusted to reflect the reality and the emotional challenges that many face when making decisions about these transitions. Expectations about what will happen influence how these transitions occur and how quickly people adjust. Preparation before, during, and after such a move is key, and as has already been mentioned, communication of what one should expect is a good tool for supporting adjustment (Chapters 7 and 8). A smooth transition is more likely with the support of staff at the new residence and access to appropriate services and support. This is also true of the final transition to end-of-life care (Chapter 9). It is argued that people need to understand that dementia is a terminal health condition, even if most people with the disease actually die from something else. Understanding and expecting decline and, eventually, death will help people with dementia and families readjust how they live with the present and better prepare for dying.

Those who communicated their expectations best among family members more successfully engaged in the transition, whether the transition was a need for outside help or a move to a residence. As discussed by most chapter authors in this book, expectations may differ between the person with dementia and across care partners, something that service providers must attend to. Finally, one solution for helping families deal with hopes and expectations is to create new knowledge about these trajectories. Robitaille (Chapter 8) makes a case for longitudinal methods such as growth mixture modeling to help us better understand what different trajectories exist and what factors influence future realities. Given the known heterogeneity of families and individuals, the more we know about these different trajectories, the more people with dementia, families, and service providers will be able to identify what can be reasonably expected for people with dementia.

Autonomy

The third message across chapters in this book is that service providers and families should not take complete control but rather ensure that the person with dementia has some substantive influence over decisional processes involved in transitions, thereby allowing them to maintain their sense of personhood.

> *Interviewer: Her finances.*
>
> *Carer: Her finances, well she's still technically in control of all of that but <sister> pays bills and stuff like that.*
>
> *Interviewer: Okay, alright. But she can still make change though eh, like if she goes out to Red Lobster or stuff like that, she can pay herself and all that kind of thing.*
>
> *Carer: Oh yeah we make sure that she has enough money in her purse and stuff like that and I have access to look at the accounts and make sure nothing untoward is happening.*

After a diagnosis of dementia, certain preconceived notions may emerge regarding the intensity and speed with which cognitive decline will occur. In terms of competency, individuals with dementia may be given the benefit of the doubt until the decline is significant enough to impact daily functioning. Acknowledging the presence of dementia may elicit biases and negative expectations, for instance, from healthcare providers or family members, that the person cannot make decisions. There is evidence that individuals with dementia can and should be allowed to express their wants and desires and their perceptions regarding their experiences, yet even family members can perceive them as being unable to do so in the moment [7]. Despite communicative and cognitive difficulties, people living with dementia can express how their dementia and other health conditions are affecting them, empowering them to take an active role in their care [8]. This is valuable information to both families and service providers. In the early stages of the disease, as they continue in their family roles (e.g., father, mother, spouse), people may desire to continue as protectors and care givers rather than care receivers. They need to be given the chance to engage as such and be given some power over the decisions that will affect them directly. It is suggested (Chapter 9) that discussions regarding advance care planning be made as early as possible, certainly while the person with dementia is still able to speak.

As noted in Chapter 2, the amount of autonomy an individual has over given situations can be determined by cultural and ethnic traditions as well as health imperatives. Certainly in Western cultures, the ability to maintain some control over decisions about ourselves is important. People with dementia, especially in the early and even in the moderate stages, can participate in conversations about driving, alternate care, alternate living arrangements, and financial control. Not allowing the person with dementia some level of autonomy over financial affairs can have disastrous results (Chapter 4) on their self-esteem. For example, going from full control of the finances to none at all is not recommended and is likely to elicit anger and frustration.

Relegating a person with dementia to someone who is at the mercy of service providers has proven to be inappropriate in hospital settings as well (Chapter 5). There is very little consideration of persons with dementia as possible candidates for rehabilitation, regardless of their current state of functioning. Yet others have suggested, for instance, that rehabilitation is possible for people with dementia if some adjustments are made to the approach [9,10]. Rehabilitation requires some level of active participation from the individual. This, of course, presupposes that the service provider views the person with dementia as capable of effecting some change, and therein lies the bias.

In this age of patient-centered and relationship-centered care, it is no longer acceptable to view individuals with dementia as helpless. Throughout the journey, professionals and families alike must ask themselves if there is a way to include the person with dementia in the decision-making process. Up until recently, very little attention was given to the inclusion of people with dementia in decision-making. In a review of findings, Larsson and Österholm [11] found that persons with dementia were most often excluded from decisions where families most commonly did not even communicate the decisions to their loved ones. In other situations, they were told after the fact or families made decisions based on their memory and understanding of prior wishes. By taking away what autonomy the person with dementia has left, we are diminishing that person's ability to be resilient and adapt to changes that come with the transition. The loss of autonomy is particularly diminished once the person with dementia relocates to another place of residence (Chapter 7). In fact, as described in Chapter 7, the more the person with dementia is involved in the decision-making, the easier the relocation will be. This seems to be the case in all of the transitions. Service providers need to resist paternalism, assuming what the person with dementia needs most, without first trying to consult them.

The desire to remain as autonomous as possible is an important characteristic of carers as they transition to first requiring help from outside agencies. A threat to independence is felt by carers as they come to the realization that they are no longer able to provide all the care needed by the person with dementia. It is important for service providers to help carers understand that asking for help does not reduce their autonomy, even if the carer becomes dependent on outside help. One can become dependent on the services of others, while remaining responsible for governing one's own life.

Dementia-friendly environments

Faced with a chronic disorder that has no known cure, persons with dementia rely on environments that are dementia-friendly to participate fully in their life activities.

> *Interviewer: Did she have trouble? I mean obviously one bedroom and then sort of sitting room and that type of thing ... was that at all problematic for her in the sense that she couldn't find things?*
>
> *Carer: Well she's always losing things, but she was doing that before when she was in a studio.*
>
> *Interviewer: So it's not so much concentrated on size or anything.*

Carer: Right, the biggest problem with the change of rooms is even though it's a smaller room, access from the bed to the bathroom is not as convenient as the larger facility. The larger room so, and that's a problem for her because she does have …

Interviewer: The incontinence and that kind of thing?

Carer: The incontinence.

Interviewer: Mmhmm, Okay, so on this floor, will she have as much assistance as she needs for her activities of daily living, her bathing and dressing and whatnot?

Carer: Yup, she'll have a little more than she's currently getting.

Interviewer: Okay, so there will be more assistance. And she's quite open to the assistance is she?

Carer: Yes, and no. She recognizes the need for it sometimes but if somebody offers to help her to do something that she feels she is capable of doing, without help, she resents it.

Interviewer: Is she actually capable of doing those things?

Carer: Yes, well sometimes, day to day it varies.

Dementia-friendly environments are designed to foster social inclusion, challenge stigma, improve services, and increase awareness [12]. To create dementia-friendly environments, we must first (and foremost) include the opinions of those living with dementia. We need open-minded societies that see people with dementia as having rights to take part in daily activities. We need organizations that are ready to change their designs and ways of doing things, and we need intersectoral partnerships.

Adapting both social and physical environments to respond to the needs of persons with dementia will allow them to take the best advantage of their surroundings, despite their functional limitations. This is especially evident in the literature on relocation to residences and long-term care. In both Chapter 7 and Chapter 8, there is evidence that the environment should be as inclusive and welcoming as possible, thereby facilitating the move to alternate living spaces. Without dementia-friendly environments, organizations such as residences and long-term care homes can contribute to prolonged adaptation periods, which increase the chances of persons with dementia developing behaviors that are generally considered undesirable. Triggers for these behaviors are now often seen as the result of stressors in the environment. Healthcare providers are encouraged to listen to care partners and persons with dementia to see which triggers in the environment are more likely to elicit increased anxiety and fear in persons with dementia. The failure to create dementia-friendly environments was also shown as contributing to problems in the transition to hospitalization (Chapter 5). Hospitals would do well by becoming more dementia-friendly so as to diminish the triggers for behaviors, reduce anxiety, and encourage full participation. For instance, knowing that noise can contribute to confusion, hospitals could consider quiet zones for patient consultations. As with a wheelchair that needs a ramp to access a building, so do individuals with communicative and cognitive disorders need a ramp to help them communicate [13].

Judicious changes

Dementia is a slow, progressively deteriorating condition, and the changes that occur are often subtle. Adaptations to these changes, then, should also be subtle and grounded in what is being experienced at the time.

Quote 1

Carer: Well not right away, no, because after speaking with the staff and all the caregivers, she's still Okay where she is, she's still safe. And she has enough friends and enough people that are aware of her and can kind of roll her back to her room, and make sure that things are all right. But they said that they would let me know, and I actually, every now and again I talk to some of the ladies there, that are a little more with it than some of the others. And uh, I say, you know, please don't hesitate, if you notice a change in Mom, just let me know. And uh, I think I'll see it coming and then we'll decide.

Quote 2

Carer: Yea, about that … That's one of the things too. What about Dad? I said, you know, if we had a routine. We don't have a routine. We don't have, and it's just too many changes for Dad. It's too much for me. I'm between a rock and a hard place.

It is very disconcerting to see a loved one's cognitive abilities deteriorate, and it is not surprising that some families might approach the changes with semicatastrophic reactions. Despite new knowledge regarding different trajectories, transitions, and outcomes, individuals living with dementia will likely follow different and personalized courses, based on their own preferences and circumstances. When faced with challenges that appear overwhelming, and armed with knowledge of what may happen in the future, families might be counseled and supported to deal with changes a little at a time. A good example of this approach is palliative care. In Chapter 9, there is reference to the fact that there can be a "dose of palliation" to ease the transition at different stages of the journey. If issues are dealt with in a judicial way as they occur, crises are less likely. The help that is offered can mirror the needs of the carer and the person with dementia. For example, members of dyads might not be ready for the next transition at the same time. Similarly, at times this disparity exists between families and healthcare providers. In Chapter 7, there is an example where the healthcare provider recommended long-term care, but the family correctly decided on a lower level of care. Adaptation to change can indeed happen in incremental fashion.

It is important that we do what we can to help persons with dementia and their families ease into the transition. What we have learned from the chapters in this book is that there are many different ways to transition from functioning "within the norms" to total loss of ability. The most successful transitions seem to include some level of adjustment to the loss; some appraisal of the types of functions and abilities that are lost and those that are still preserved. Even at diagnosis (Chapter 2), there are opportunities to connect families and individuals with dementia to organizations and programs that offer an array of services that can be accessed pending their needs (e.g., First Link programs in Canada). The timing of what is needed and when is an important factor to consider in smoothing transitions.

Understanding and creating new aspects of self

Part of the experience of dementia is, on the one hand, letting go of some aspects of who the person with dementia once was, while at the same time preserving other familiar aspects of their person. What we have learned from the research so far is that successful transitions include accepting the new aspects of the person with dementia. Change might be thought of as an entirely normal human characteristic. We all do it. Expecting change reduces the threat of it and allows us to manage, often through changing ourselves.

Quote 1

Carer: The openness and the relationship. Um, he's not diminished in my eyes, he is not. But, we still, I still enjoy living with him, and that's you know, I can't imagine. Well, I can't imagine anything yet that would spoil that.

Quote 2

Carer: Absolutely. Everybody agrees, because she is very socially engaged. Yes, she is suffering memory issues and doesn't have the same reasoning skills that she once had. But other than that, her emotional state is actually, in the company of people, very positive and engaging. She is a happy person.

Some carers have described the experience of dementia as a funeral that never ends. Families and friends witness their loved ones change before their eyes in ways they never expected. A very vibrant, active, intelligent man can deteriorate to the point of not responding to his name or being unable to feed himself. It is indeed a grieving process that brings out the most human side of all of us. The grief, of course, is related to the person the families feel they have lost, and coping can be difficult as they are faced with the deterioration of the self. So how can individuals and families hold on to the aspects of the person that was all the while accepting the new aspects of the person?

In Chapter 6, we have found that the families that seem to find the transitions the smoothest are those who have accepted that the person with the dementia is changed. But it is not the entire person that has changed. Families struggle when they relate to healthcare professionals who only see a characterization of the person with dementia. As families see the service providers interact with the "new" person with dementia, they attempt to have the person they have known recognized for who he or she used to be and for what their loved one could do in the past. A lot has been written about the important consideration of personhood in dementia [14]. One strategy that is often used to connect health service providers with persons with dementia is to share biographical information concerning the person with dementia. But unfortunately, service providers do not always take the time to find out who the person is that comes to consult or ask for services, and practice shows that files are seldom read [15]. So how can one reconcile the need to let go of some aspects of the old self while identifying with some aspects of the new one?

One possible explanation for understanding how this is possible is to conceive of *the* self as a multifaceted entity. Neisser [16] suggested that the self is actually made up of several types of selves that derive their information from different types of environmental cues. For instance, the ecological self would derive information from interacting with the physical environment. This information would be processed through the senses, whereas the interpersonal self would derive information based on social communication. Baird [17] later adapted these notions to dementia experiences, using one of her clinical cases. In describing the person with dementia, she supports the view that we indeed have several "selves" that are not simultaneously changed with dementia and that in fact some of these can be preserved and others destroyed. In moving from transition to transition, then, it might be best to understand how people living with dementia might focus less on their *conceptual self* as "a concept of oneself as a particular person, for example, with social roles and social/cultural differences" (p. 2), while preserving other aspects of their selves. Families might also do best to shift their conceptions of the person with dementia by emphasizing other aspects of their selves as they change throughout a transition.

For instance, the *interpersonal self,* "the self engaged in personal interactions" may not be impacted by the dementia until the very last stages of the condition. By understanding that we might consist of more than oneself, we could come to terms with the fact that the person with dementia has lost some parts of their self while preserving others. Perhaps the best way to live through transitions as carers is to focus on those aspects of the other self that are most adaptive to the transition, as well as on those aspects that remain constant: letting go the former, maintaining the latter, reevaluating, and readjusting.

Individuals with dementia can also come to terms with the fact that some parts of themselves have changed but that others have not. Healthcare providers would do well in helping individuals capitalize on what is preserved. Especially in long-term care situations, understanding that there is not only one self may help staff understand that other aspects are preserved. They would benefit from enhancing the expression of these aspects. Focusing on strategies that enhance some understanding of the other parts of the self may create strong resilience as the person transforms. Providers must become familiar with the person with dementia in their care. Keeping staff consistent maximizes the chances that the focus during transitions will be on those parts of the self that are preserved. Families are frustrated at having to constantly update new service providers about aspects of the "new" and "old" person with dementia. Chapter 9 brings these points to the end of life by focusing on aspects of change as an individual approaches death.

Savor

We often speak of the dementia experience as something that we are subject to or that we have to endure. This may be a realistic characterization of the journey, but another way to conceive of the experience is to savor moments as they are encountered.

Interviewer: Okay, so what do you expect from the move then? You're expecting her to enjoy it?

Carer: Well exactly, to enjoy and to not to feel so lonely and I guess the other thing is not to worry as much. Because I know for a fact that there is someone in the morning that goes to give her medication, and over all they keep an eye on their residents. It is comforting to know that they are keeping an eye on her and that she is not as lonely. She may be lonely in the evenings but there are activities at night and socials so she is meeting people, she is friendly.

In Bryant and Veroff's 2017 book called *Savoring: A new model of experience* [18], the authors suggest ways for how we might characterize events in a positive light, despite the existence of negative occurrences. In the area of dementia, the writings so far have brought attention to the challenges faced by people with dementia and their loved ones. There is no doubt that focusing on these aspects has brought attention to very real obstacles and has led to changes in services and policy. However, a focus on the negative may not be the best recipe to help those with lower resilience, who fail to grow when faced with negative events. The concept of savoring involves bringing positive emotions to the surface, thereby allowing individuals to focus on the enjoyment. Somewhat similar to our discussion about parts of the self, families might be better served by acknowledging that the person with dementia has been stripped of some aspects of enjoyment but that, at the same time, they must not ignore the parts that bring joy. The concept of "savoring" focuses on our reactions to experiences, and Bryant and Veroff [18] outline three primary elements to this process. As with other psychological processes, savoring involves (1) experiencing the sensations, (2) focusing on the feelings of the experience, which in this case should involve mental processes and operations that help individuals focus on the positive, and (3) developing concrete strategies and actions that impact positively on the event.

Counseling people living with dementia and their carers about how to savor experiences may help them develop the resilience needed to transform their emotional and behavioral responses to reflect positive interpretations of those experiences. Smith and Hollinger-Smith [19] found that higher savoring scores actually predicted higher levels of happiness in a sample of 164 older adults. This result was significant for individuals with both higher and lower resilience scores. Resilience allows us to continue to develop in the face of negative events. Resilient people will flourish, but this may take time. Sabat [20] eloquently describes the case of a spouse who, over time, learned to adjust her emotional reactions to the changes she experienced with her husband and, indeed, even succeeded in flourishing as she supported him during his stay in long-term care home. Savoring has to deal with the attention we place on positive emotion. It is not surprising that these concepts are related and impact our sense of happiness. The conclusions drawn from the previous chapters in this book point to strategies that may indeed help individuals with dementia and their carers not only learn to ride the waves but also maintain a certain level of happiness if they learn to savor the moments.

Team of helpers

The final advice for individuals with dementia and their families is to help them come to the realization that they typically cannot ride the waves without help. Clinicians and policymakers need to offer programs that will indeed offer support and be appropriate for various stages of the journey.

> *Carer: That's what I'm trying to do. Like that's why they try to give us the 8 hours and the lady is coming 2 and a half hours a week. We can drop, cut it down to 2 hours, the half hour that we usually pay her per day. We can probably use that, hopefully, to go once a week or twice a week or something. Whatever we can get, because I know he's bored at home.*
>
> *Interviewer: No, if there's something you could find, that'd be an excellent idea.*
>
> *Carer: Especially now that he has Para-Transpo, that's so helpful. Because he can make arrangements with Para-Transpo to come pick him up. And I buy tickets for him and he can go. And they can pick him up and drop off. Right now, we're just very lucky, the family, that I work shift-work, at this point. I don't want to, if I can find a regular 9–5, my dad's going to be, we're going to have to share it, because you can't take time off work all the time.*

It is important for carers to access outside services to care for themselves and others. This means that services must be made available to families of all cultural backgrounds and languages, that there is a realization that the types of teams of helpers and the nature of the help will change over time, and that there needs to be support to help families deal with the stigma related to requesting help. A great deal has been written about support for carers and the importance of taking care of themselves. As mentioned in Chapter 6, caregiving can have devastating impacts on all fronts: physical, psychological, financial, and social. Families, friends, and others should try to work together as a team to reduce these overwhelming impacts, and this may entail collectively identifying what is most stressful and how to best readjust to the changes. Good communication and overt respect among all care partners, and between care partners and service providers, will best achieve equilibrium to complete the transition. As stated in Chapter 1, trust is an important feature of these relationships. Without this trust, the persons with dementia and their carers will continue to bear the burden of responsibility. This is true even after death. After having embraced their roles of carer, families need to relinquish these roles. Support is important for carers as they turn to both social and professional connections without the person with dementia.

Conclusion

Dementia is a big part of the lives of an increasing number of individuals worldwide. While the international scientific community works hard at characterizing the different dementia trajectories and finding cures for the health conditions that cause dementia, the world is left with an increasing number of individuals who must currently live with the challenges that will be presented to them. We would like to believe that clinicians, policy-makers, and researchers, working with the help of nongovernmental agencies, people with dementia themselves, and carers, have begun to change the discourse on dementia.

We have learned through this book that the trajectories will be different for everyone, that the transitions are real, and that most people with dementia will probably go through at least some of them. We have learned that reactions to the challenges can vary not only by intensity but also by how individuals react to each wave. Some find it tough at first and then find strategies to adapt to even larger waves, while others find it easy at first and then struggle as the person with dementia loses some of the parts of their self or becomes increasingly unable to engage in functional activities. We have learned that our reactions to these events are important and that individuals with dementia and their families will be better served if strategies are developed to adjust preferences and expectations, develop care in manageable fragments, seek help, and focus on the positive.

We have introduced Kopsov's model [5] as it accounts for individual needs and it summarizes subjective well-being as a dynamic process that is influenced by needs, knowledge, experience, and choice. All are capable of influencing the experience of dementia and transitions and can serve to support us to "readjust" as we ride the waves. The more we learn about the experiences of people with dementia and the more we learn to properly characterize these complex trajectories, the more we will be able to serve and satisfy their needs.

Where next?

There have been many common messages throughout this book. The best advice, and interventions that service providers can give to help families living with dementia, will be sustained by a deep understanding of the experience from the individual's and carers' perspective. This book has been a discourse between science and "lived experience," hoping to inform both.

Good, timely information will contribute not only to the identification of the person's needs but also to the possibility of these needs being met. This begs for better and more efficient public awareness campaigns, not only about diagnosis and the early signs and symptoms but also about all transitions, including information about advanced care planning.

These messages should also carry some element of positive information, helping families find hope and joy, so they may savor the good moments that are ahead. This information needs to be communicated publicly, but equally important is the communication of this information on a one-on-one basis between service providers and people with dementia and their carers.

Communication between care providers and families presupposes some understanding of the heterogeneity of this population and some understanding of the particular needs of populations who are from different ethnic backgrounds, geographical locations, or genders. Finally, research into the impact and accessibility of services to these groups will go a long way in offering a series of options upon which families can act. More research using longitudinal approaches will contribute greatly in examining the trajectories of individuals and groups with similar mitigating factors. Not only will this lead to better policy design but will also help inform families and people with dementia about the likelihood of events occurring for their particular case.

In closing, it is our duty as a society to help those who struggle with the consequences of significant chronic health conditions such as dementia to live the best lives possible, to find safe places to live, to find meaningfulness in their lives regardless of their impairment, and to be surrounded by loved ones. While we might be riding the waves on our own, it behooves us to offer and seek professional and informal help to assist those who are challenged by them. Clearly, the waves of dementia-related experience do not give the lone, unknowledgeable, inexperienced paddler an enjoyable ride.

References

[1] Wilkinson AM, Lynn J. Caregiving for advanced chronic illness patients. Tech Reg Anesth Pain Manag 2005;9(3):122–32.
[2] Wilson I, McDermott H, Munir F, Hogervorst E. Injuries, ill-health and fatalities in white water rafting and white water paddling. Sport Med January 14, 2013;43(1):65–75.
[3] Chertkow H, Feldman HH, Jacova C, Massoud F. Definitions of dementia and predementia states in Alzheimer's disease and vascular cognitive impairment: consensus from the Canadian conference on diagnosis of dementia. Alzheimers Res Ther July 8, 2013;5(Suppl. 1).
[4] Drummond N, McCleary L, Garcia L, McGilton K, Molnar F, Dalziel W, et al. Assessing determinants of perceived quality in transitions for people with dementia: a prospective observational study. Can Geriatr J 2019;22(1):13–22.
[5] Kopsov I. A new model of subjective well-being. The Open Psychol J 2019;12:102–15.
[6] Clissett P, Porock D, Harwood RH, Gladman JRF. The challenges of achieving person-centred care in acute hospitals: a qualitative study of people with dementia and their families. Int J Nurs Stud 2013;50:1495–503.
[7] Miller LM, Whitlatch CJ, Lyons KS. Shared decision-making in dementia: a review of patient and family carer involvement. Dementia 2016;15(5):1141–57.
[8] Whitfield K, Wismer S. Inclusivity and dementia: health services planning with individuals with dementia. Health Policy 2006;1(2):120–34.
[9] Clare L. Rehabilitation for people living with dementia: a practical framework of positive support. PLOS Medicine 2017;14(3).
[10] Maki Y, Hattori H. Rehabilitative support for persons with dementia and their families to acquire self-management attitude and improve social cognition and sense of cognitive empathy. Geriatrics 2019;4(1):26.
[11] Taghizadeh Larsson A, Österholm JH. How are decisions on care services for people with dementia made and experienced? A systematic review and qualitative synthesis of recent empirical findings. Int Psychogeriatr 2014;26(11):1849–62.
[12] Alzheimer Disease International. Dementia Friendly Communities. 2019.
[13] Kagan A. Supported conversation for adults with aphasia: methods and resources for training conversation partners. Aphasiology 1998;12(9):816–30.
[14] Hennelly N, Cooney A, Houghton C, O'shea E. The experiences and perceptions of personhood for people living with dementia: a qualitative evidence synthesis protocol [version 1; peer review: 2 approved]. HRB Open Research 2018;1:18.
[15] Alzheimer Disease International. World Alzheimer Report 2018 - The state of the art of dementia research: New frontiers. 2019.
[16] Neisser U. Five kinds of self-knowledge. Philosophical Psychology 1988;1(1):35–59.
[17] Baird A, Completed. A reflection on the complexity of the self in severe dementia. Cogent Psychology 2019;6:1574055.

[18] Bryant FB, Veroff J. Savoring: A New Model of Positive Experience. Mahwah, New Jersey: Lawrence Erlbaum Associates, Inc; 2007.
[19] Smith JL, Hollinger-Smith L. Savoring, resilience, and psychological well-being in older adults. Aging & Mental Health 2015;19(3):192–200.
[20] Sabat SR. Flourishing of the self while caregiving for a person with Dementia: A case study of education, counseling, and psychosocial support via email. Dementia 2010;10(1):81–97.

Index

Note: 'Page numbers followed by "t" indicate tables, "f" indicate figures and "b" indicate boxes'.

M

N

O

P

Printed in the United States
By Bookmasters